BLUEPRINTS
FOR THE RESIDENTIAL CONTRACTOR

BY GEORGE PAPAHERAKLIS

Contact the authors or publisher if you would like information on how to access any and all programs or other materials associated with this book and its contents.

All photos courtesy of Soleimani Photography.

ISBN-
First Edition

NOTE
FROM THE AUTHOR

I've written this book to help the entrepreneurial contractor who wants to take the shortcut to owning his contracting business instead of having his business own him. These pages represent the tried and true successful actions that have gotten me from the barest beginnings to what people consider the influential top of my game. With more than 40 years of experience, being an award-winning contractor and having serviced well over 1,000 homeowners, I've reached a level of experience, knowledge, and wisdom that I want to give you.

This book contains simple but powerful and proven truths which, when applied, have exact outcomes. They will help you in your journey and they will bring you success. This workbook will give you all the tools you need to transform your business and your life. Make notes, use the documents we've provided, and make all of our ideas fit into your unique business.

I'm available for you, so if you want to contact me, shoot me an email at customercare@finecraftcontractors.com.

Enjoy the book! After all, it's for you.

TO YOUR SUCCESS!
George Papaheraklis, B.Arch.
Founder & President of FineCraft Contractors, Inc.

PROLOGUE
FROM THE EDITOR

IN DECEMBER 2017, THE PAPAHERAKLIS FAMILY invited me into their business and their lives. They'd hired me to write a book about everything their patriarch, George Papaheraklis, has learned and all of the practices he's honed after more than 30 years as the leader of their business, FineCraft Building Contractors.

I HAD NO IDEA WHAT TO EXPECT.

I arrived on a Sunday evening, the night of FineCraft's Christmas party for customers and colleagues. After a harrowing Uber ride, I was dropped off at a beautiful home in Bethesda, Maryland—a home which FineCraft had renovated for a lovely couple and their two kids. I had only ever spoken to Niko Papaheraklis, George's son, over the phone, so I didn't even really know who to look for.

When I entered the foyer, I explained to a number of different people wearing FineCraft buttons who I was and why I was there. Niko popped in, introduced himself, and then popped back out to continue getting ready for the party. Someone thrust a FineCraft button into my hand, which read "I'm Family." I promptly attached it to my shirt and then, there I stood, a representative of FineCraft whom no one knew. I stuck out like a sore thumb, if I'm being honest, and stood around wondering how I was going to convince these people to talk to me and tell me about their business.

It took all of fifteen minutes for this family's natural hospitality and passion about their business to overcome my own nerves and I was soon talking and laughing and learning all about them and their business. And just as quickly, I got a sense of why they were so successful.

That night was a parade of FineCraft customers and colleagues, each of whom was greeted personally by every single member of the Papaheraklis/FineCraft family. It was clear they all knew each other, exchanging stories and asking after each other's families.

Everyone I talked to that evening had nothing but great things to say about FineCraft, about their work ethic, integrity, honesty, and the quality of their craftsmanship. I left that party with a clear sense that FineCraft's success has just as much to do with the Papaheraklis family as it does with their phenomenal work.

I would spend the next three days riding around with George in his truck, an opportunity to be a fly on the wall and observe first-hand how he runs his business. I listened as he switched seamlessly from Greek to English to Spanish. I visited job sites from Washington, DC to Chevy Chase to Bethesda, Maryland to Virginia. I listened as George and his right-hand-woman, ViVi, addressed customer concerns, picked up supplies, calmed stressed employees, fixed inadequate workmanship, and so much more.

EVERYONE I TALKED TO THAT EVENING HAD NOTHING BUT GREAT THINGS TO SAY ABOUT FINECRAFT...

I also spent time at George's home, meeting his entire family, including his wife, Dagmar, granddaughter and great-nephew, and observing as this loud, caring, and funny family (a large majority of whom live in one block of a quiet neighborhood in Gaithersburg, Maryland) navigate life and business together.

THIS BOOK IS WRITTEN FROM GEORGE'S PERSPECTIVE. He's the expert. These are all his words, his experience, his expertise. But as I began collecting my thoughts and writing down everything I'd learned from George and his family in the days I spent with them, I realized that George couldn't necessarily speak to certain things that seem to make him successful. He wasn't listening as I spoke with past and current clients at the Christmas party or as I observed his interactions with clients and how he could immediately waylay their concerns about a project or give them the cold, hard truth about a wayward timeline.

So I decided that I would place my observations and the observations of those who know George best throughout the book. You'll see these observations, labeled **EDITOR'S NOTES.** Hopefully, these give you a sneak peek into how George handles the day-to-day grind of running a demanding business.

KRISTINA BRUNE
Editor

4

INTRODUCTION
BY NIKO PAPAHERAKLIS

As a child, I thought in order to make money one must work hard—really hard, as in waking up early and going to bed late, spending minimal time with your family. I thought it was simply a given that the father would be the breadwinner of the household and it was his duty and obligation to burn the midnight oil for his family's wellbeing.

No one told me this explicitly. I came to believe this because I never saw my father. Every morning he would be gone before I walked downstairs and I would be asleep before he got home.

I remember hearing him walk downstairs from his bedroom every morning—it was 13 steps down. The first thing he'd do was walk outside and start his truck. Then he would come back inside, prepare his coffee, and make two pieces of toast with cheese and jam. He placed the toast one on top of the other, separated only by a paper towel, on top of his coffee mug. Then he was off for the whole day and much of the night.

We never talked much about how little time he spent with us. We simply understood that he was making money so we could have everything we needed. However, that didn't stop me from wishing I could see my father in the mornings or simply wanting him to be more involved in our lives instead of apart from our lives.

As I got older, I decided I wanted to taste what a day was like in the shoes of my father. So I went out with him one cold winter day. My mom packed us some food, ensured I was dressed for the occasion, and we were out the door by 6:00 a.m.

I walked outside that morning and headed toward my father's truck—a blue 1985 Ford van with manual transmission. It had hundreds of thousands of miles on it and was so old that he practically had to hit it before it would start. He definitely had to make several attempts before it actually started. I remember feeling embarrassed.

In my mind, my father was this God who was providing us with food, money for school, and a roof over our heads. My father should have had the Rolls Royce of trucks. I couldn't understand why he didn't have a better mode of transportation even though he worked harder than anyone I knew.

When we arrived at the first job site, he was like a general who controlled his guys like soldiers preparing for war. As soon as he arrived they woke up and started moving faster—their shuffle turned to a sprint and idle hands got to work. His arrival was like their morning dose of caffeine.

I remember many things from that day, but what surprised me the most was that he worked harder than anyone on the job. He knew how to position the guys so they would get the most done in the least amount of time. He'd get as involved as needed, doing the work needed to get the job done, no matter what it took. If he saw something that needed to be done, he'd either have someone handle it or he'd handle it himself. If there was a problem, it would be fixed. It didn't matter if the client saw it or not—if it wasn't up to his standards it would be fixed or redone.

This is what I came to understand about my father that day. He leaves no stone unturned. Quality is in every single nook and cranny of every single project he touches. He will sacrifice anything—even time with his family on a Sunday morning—to ensure his projects adhere to that quality and his clients are happy.

What I didn't understand was if my father worked so hard and had such high standards, why was he always away from his family? Shouldn't he be making enough money by now that he could relax a bit? It didn't make sense that he would have to be so overworked.

Years went by and my father's schedule didn't change. His work controlled his life and he lived to work. From a son's point of view it was quite hard to have a father but not have a father at the same time. I struggled to make sense of his absence. I couldn't understand what the problem was, so I arrived at the simplest answer: My father was a workaholic. That was good enough for me. A workaholic works all the time, right? It was the easiest explanation.

As the years wore on, it became harder and harder to accept. So difficult, in fact, that I started to dislike my father. I had nothing in common with him. To add fuel to the fire, it dawned on me that our schedules were actually very similar, except for one small detail—my day had predefined limits and his didn't. My day was defined by a schedule. His day was defined by production.

This brought up a very interesting question that I wanted an answer to. Why couldn't my father produce within a predefined schedule so he'd have more time with his family? What had to change to make this happen? For God's sake, we all lived by the same amount of time each day—shouldn't it be a luxury of getting older that one is more in control of his life?

Well, it's been a long, hard road, but I'm happy to report that today, my father has changed the way he works. He starts and finishes his day at a reasonable hour. He spends more time with his family. He doesn't do it all by himself anymore, and as a company, we get a lot more done and finish ten to twenty times more projects annually than back in the eighties, when my father first started FineCraft.

My father built our company, FineCraft Building Contractors, from the ground up. Throughout my life, I witnessed both his successes and failures, all of which ultimately led to the successful business we run today. What I'm most proud of is that he's been able to find a balance between his personal and professional life.

If you recognize yourself, your family, and your schedule in these pages, you are not alone. Business owners, especially those of us in the contracting industry, too often feel like they have to handle every aspect of their business in order to have success, no matter the cost to their personal life.

But we—me, my father, and our FineCraft family—are here to tell you, *it does not have to be that way.*

In the pages of this book, you will find a guide to regaining control of your life and growing your business. You can achieve business success and be present with your family, enjoy a jog or a bike ride in the middle of a workweek, or even go home and take a nap. In other words, you can have a life you're in charge of.

NIKO PAPAHERAKLIS
Son of George Papaheraklis
Business Manager of FineCraft

TABLE OF CONTENTS

CHAPTER 1
MY STORY

I LIKE TO SAY THAT I'VE HAD THE PRIVILEGE of experiencing 400 years of evolution in my lifetime. It's a life that prepared me for hard work and success.

I was born in the mountains of Greece. The village I was born in had an elevation of about 4,500 feet. It was a very remote and relatively unfriendly place to live. In winter there was usually about six feet of snow from October to the middle of March or April. There was no running water, no electricity, no roads. The town had one telephone line you could use to call to the nearest little town. Carving out a life there was a toil for survival from the beginning of the year until the end. Being born into those conditions gave me a mindset that you have to work for survival the whole year.

At four years old, I saw my first bulldozer. I could hear it coming before I saw it. When it came around the side of the mountain, it seemed like a monster. For the next year I followed that bulldozer around as the workers tore into the earth with machines and dynamite. I watched them set the dynamite and saw the rock rain down like confetti. To this day, the smell of kerosene and diesel brings me right back to those moments.

I built my first building when I was four or five. Many people in the village erected miniature chapels for their favorite saints, so, using rocks, I started building them three or four feet high in the fields that surrounded our village. I was fascinated by the idea that I could build something with my own hands. I think they're probably still standing there in Greece, all these years later.

At six years old my father taught me how to draw. I was very good at it. I learned to love art at a very early age and before my family left Greece, I won several awards for painting. Had it not been for some turn of events when I came to America, I would have dedicated my life to painting.

When I was nine, we moved from the mountainside to Athens and my siblings and I entered school. I was a great student, even though the education system in Greece was very rigorous. At the junior high level we would take 10-15 subjects including art, geography, literature, math, history, world history, Greek history, biology, ancient Greek, Latin, music, science, and more.

I excelled in math. I was so good, in fact, that the teacher would often ask me to teach the other students. Which, of course, put me in a terrible position with the other students, who hated me and made my life miserable between classes.

In addition to my schoolwork, I helped my father in his store. He was working 18 hours a day and because I was the oldest of three children, it fell on me to help after school.

In 1969, when I was 15, my uncle helped my family immigrate to America. I was dumped into the United States' education system two days after we arrived. We came on a Saturday and on Monday morning, I attended my first day of school. I didn't speak one word of English.

This school was so different from my school in Athens. The school was 10 times as big and students traveled from classroom to classroom, whereas in Greece, we had our own classroom for the whole year and the teachers came to us. They also only studied five or six subjects and the requirements seemed much more relaxed than in Greece.

Because I didn't speak a word of English, I just didn't speak at all at school for six months. I was able to participate in algebra a bit, because I could simply answer the questions on the board. But most of the teachers assumed I was disabled in some way. I would communicate as best I could via hand signals and they would just pat me on the back.

I was determined to learn English. Armed with only an English to Greek dictionary, I would go home after school each day and pick up anything with words—newspapers, magazines, anything, and I looked up the definition of every word.

In six months I was able to speak English fluently. The first time I felt comfortable enough to speak at school was with my biology professor. I simply went up to his desk and said, "Mr. Perora, I want to ask you a question."

He was so shocked, he said, "The idiot speaks! It's a miracle!"

I replied, "I speak modern Greek, common Greek, ancient Greek, and I now speak English. How many languages do you speak?" He apologized and from then on, my time at school was very different. I excelled in my studies and showed those teachers that I was not an idiot.

My uncle who had helped us immigrate owned a restaurant in Maryland. I learned early in my life that I don't like having debts and I felt indebted to my uncle for helping us. So I worked evenings and weekends for him in his restaurant for a year for almost no pay. When I felt my debt was paid, I quit.

After that, I went to work for a local carpenter. There are certain people I owe a lot to, and this man is one of them. He was a lovely guy, a big, burly hillbilly with a beard. I learned by observing him and his work ethic. He not only taught me about contracting and carpentry, but he taught me the value of competence. After working for him for a couple of summers, I decided to start my own business and became my own contractor, which I would do all through college.

I EXCELLED IN MY STUDIES AND SHOWED THOSE TEACHERS THAT I WAS NOT AN IDIOT.

I finished high school with a 4.0 average, number 17 out of 1,500 students. I got a scholarship to go to college. My father decided that since I was so good at math and art, that I should just combine the two and become an architect. Not knowing any better, I agreed.

I went to the University of Maryland and majored in architecture. I did well, however, it killed my love of drawing. Architecture puts you in a straightjacket of making straight lines. So when I started my internship and realized that they were going to have me draw bathrooms, I asked myself, "What am I doing here?"

All throughout college, I had my own construction company, doing small projects like additions and remodeling. I was making a lot of money. I decided I did not want to draw bathrooms. I wasn't going to be an architect. I realized I could combine my love of art, my excellence in math, and my architecture knowledge to do contracting differently than 99% of the contractors out there.

The rest, as they say, is history. Though I got my Bachelors in Architecture, I worked very hard, for many years, to build FineCraft into the company it is today. My family might say I worked too many hours and spent too much time away from home. My business survived the inevitable ups and downs of the contracting business. I made mistakes and I had successes. I'll talk much more in future chapters about the strategies and business practices I used to ensure success, but there are two things I learned as a young man that have been the cornerstones of my life as a contractor: competence and confidence.

My hillbilly boss taught me competence. I learned to love competence and I think competence is one of the most important things in life. If you're competent, you can get away with anything. You can be a complete jerk, but if you're competent, people will forgive you to a large degree. Conversely, you can be the most beautiful guy in the world, but if you're not competent, people will shove you around, push you down, and you're not going to get anywhere in the long run. A person who is competent makes it in life. Because of my experience with him and seeing what he could do, I strive to be competent in my life, in whatever I'm doing.

I BELIEVE IT'S ONLY THROUGH KNOWLEDGE THAT PEOPLE CAN BECOME BETTER

I also learned the value of confidence. How do you get people to trust a 20-year-old kid with a construction project? Confidence. People would ask me if I could do a railing. And even if I had never done a railing before in my life, I would say, "Yes, yes I can." My certainty was enough to show them that I could do it.

Then, if I didn't know how to build a railing, I would open up the books and learn how to do it. Then I would go and make a beautiful railing. It took me however long it took me and that was it. I never said I couldn't do something. I always bluffed my way through until I was able to actually do it. I never went in without really trying to find out how.

A FINECRAFT PROJECT FEATURED IN BETHESDA MAGAZINE

Another valuable lesson I learned on my journey is that there are two types of people in the world. One type will freely give you their knowledge. The others will hide it like it's gold. They just don't want you to know.

I had many people who helped me and I decided early in my life that I was going to do everything I possibly can to give whatever knowledge I have to anyone who wants it. I want to give knowledge, of any kind, because I believe it's only through knowledge that people can become better.

It is in this spirit that I'm writing this book. I want to share everything I've learned in the almost 45 years I've been working as a contractor, in order to help you build or grow your business.

I was able to combine my passions—art and math—along with a dedication to hard work, into a job that I love. So whether you are a contractor because it pays the bills or because you're passionate about it, I encourage you to seek out whatever drives you and maintain a focus on that passion. You will use this to push through the hard work and dedication it's going to take to be successful.

NOTE FROM THE EDITOR:

George told me the story of his childhood and how he came to create FineCraft as we sat outside on his back deck on a chilly December night. His whole family was there—his wife and children, his sister, brother, cousins. As he talked, his younger brother Pete frequently interrupted to give him trouble or clarify a fact. Ultimately, George would either prove that he was right or simply move on. He smoked a cigar, drank wine, showed me his artwork (he is indeed incredibly talented). That night, his wife and son would dance in the living room, his great-nephew would play on the iPad with me, we would listen to Greek music, Latin music (George's wife Dagmar is from Panama), and laugh a lot.

I tell you all of this to give you an idea of a man who sometimes feels a bit larger than life. He and his wife's life stories are the things of movies. But at the end of the day, he's a man who loves his family, loves his work, and really does want to tell you everything he knows.

— NOTE!

TIPS FOR SUCCESS

- **Have confidence in everything that you do. If you're not sure how to do something, go out and learn.**

- **Know the value of competence. When you're hiring employees or choosing vendors, pick people who are competent, even if they don't have the most impressive resume or the flashiest business.**

- **Freely give knowledge instead of withholding it from others. It makes the world a better place and makes us all better.**

- **Have passion in everything you do. If you don't love this work—I mean really love it—it's going to be hard to be successful.**

- **You won't be successful unless you work hard. Plain and simple. Hard work is a requirement of this job and if you're not willing to do it, you'll never make it work.**

NOTES

CHAPTER 2
THINK ABOUT CONTRACTING DIFFERENTLY

I STARTED MY BUSINESS YOUNG. I was a young man, on my own, trying to make enough money to get by. I learned early on that I could—and should—do everything on my own. After all, it was up to me to make this business work.

I carried that belief with me for a long time. And I made plenty of mistakes because of it.

For 15 years, I had a partner I never should have had. I suffered his narrow-mindedness and his deficiencies for years. In retrospect, I thought I could improve his behavior and work habits. Also, looks can be deceiving. He made me think he was important to have around. But more than anything, I gave my word that we would be partners, so it was hard to go against my word and get rid of him.

One of the many reasons I disliked working with him was his attitude towards me when I asked for help. At this time, it was just the two of us with maybe one employee. Our standard practice was that when we were working on a job site, we would divide and conquer. He would work on one project at the site and I would work on another. This way, we could get more done at one time.

However, there are many parts of the construction process that simply require more than one person. For example, one day I was preparing to hang a 10-ft long piece of joist in the ceiling and I needed to nail it up there. Any normal person with any construction experience at all would understand that that is at least a two-person job. So I called to him from across the house to help me.

Each time this happened, he had a terrible attitude and would say something like, "This is getting old. You're interrupting me so that I can help you!" This, even though he would frequently ask me for help!

On that day I'd had enough. When he let me know—loudly—that I was inconveniencing him, I looked at him and said, "You son of a bitch. Fine, I'm going to do it and I'll show you I can do everything myself." And you know what? I did it. I hung that joist. I learned how to do everything myself.

It felt good then, but it was a mistake. It was a mistake because I became so self-sufficient that I was only in my own mind. I believed I was the only one who could run my business. I didn't train anybody and I wasn't concerned with growing.

So what happened? My business didn't grow for a long time. As you read in Niko's introduction, I was away from my family constantly. I was doing everything and I was exhausted. I was working at the job sites, I was doing the accounting and ordering and marketing (what little there was of it). I wasn't training anybody and I was just barely hanging on.

Here is what I eventually learned. We will always only have two hands. No matter how much you might wish you had 100 hands, you'll always only have two. So doing everything on your own is not sustainable. There will come a time where you wish you had done something earlier—hired a crew, focused on marketing and advertising, expanded your business, whatever—and you didn't do it because you are only one person. But by then it will be too late.

Luckily for me, I figured this out before it was too late. I had a business consultant come in and look at my business. The first thing he asked me was, "How are you going to replace yourself?"

I said, "Replace myself? Why would I need to replace myself?"

"Well," he said, "you're doing everything yourself."

"Yeah and I can do it, thank you very much," I responded.

He said that wasn't the point. It took me a long time to understand what exactly his point was, but eventually I understood something. My business wasn't going to grow until I built a team around me who could take on some of the work.

This is the number one problem that every contractor out there with a van and a couple of tools is going through. This belief that we have to be everything and do everything for our business is accepted all across our industry. But hear me—it is a trap.

WE NEED TO CHANGE THE WAY WE THINK ABOUT CONTRACTING. YOU DO NOT HAVE TO DO EVERYTHING YOURSELF. YOU CAN BUILD A GREAT TEAM AROUND YOU AND GROW YOUR BUSINESS!

Here are some ways you can flip your thinking about what we accept as traditional contracting truths.

WORK FOR OTHER CONTRACTORS

I know what you might be thinking. *I'm trying to build my business. Why would I go work for someone else? Especially since I barely have enough work to pay people.* Well, you don't have enough work for them because you don't have anybody out there looking for work. There are other options for you that will allow you to build a team and find more work.

Plenty of larger contractors will hire guys and their crews to complete smaller jobs. You can work for them while you're growing. Sure, they'll take a percentage of the pay, but if you grow to be big enough, eventually you won't need to work for anyone else anymore.

In other words, stop the narrow-minded way of thinking that all of your work needs to come from homeowners and architects. You can get a lot of good, solid work from other contracting companies. It pays to market to another contractor who can give you work. They pay for the marketing and get you the job. All you need to do is deliver the goods. And man, do you need to deliver, because the work will stop coming your way if you don't deliver as promised. If you don't uphold your side of the bargain, the projects coming to you will come to an abrupt halt.

NEVER SAY "I KNOW"

You're not one of the big dogs yet. You're still growing your business. There is plenty that you can learn from those of us old timers who have been around forever.

If there's one thing I have done in my life that I will never regret, as it applies to construction, it is that I have never said, "I know." That is important because the moment you say, "I know," about something, you stop learning. Why? Because you think you know it. Therefore, you think there's nothing else to know. But if you say "I know, but I can always learn something better or more or learn a better way," you can always learn how to do something more efficiently.

A perfect example is a carpenter I once worked with. He was supposed to be the best carpenter in the history of carpentry. And he was good, but he was stuck on a particular way of doing things.

He would sharpen his chisels on a particular honing stone that came from a particular place that is the best place to get honing stones. Then he would give the door just the right bevel. Then he would take the rest of the day to install that door, carefully putting on the hinges, doing the beveling, etc.

Now, don't get me wrong—the doors were beautiful, just gorgeous. Except nobody would dream of charging clients so much for a door! It wasn't worth it. But he refused to change his system. Any kind of suggested shortcut or way to speed the process up was a huge sin in his eyes.

I decided I had to find a better way to do doors. So I sat down and I figured out a way. Through trial and error I got better and better. By the time I finished perfecting it, I was able to take 14 doors from scratch and hang and install them in one day. I also learned how to install all the cabinets in a kitchen in a day, by myself.

That carpenter lost my business because I learned how to do it more quickly. If he had been open to hearing my ideas, he could have kept my business.

FIGURE OUT WHAT YOU DON'T KNOW AND EITHER TEACH YOURSELF OR LEARN FROM SOMEONE ELSE.

I realize this seems like the opposite of what I said earlier, about not doing everything yourself. But I learned and then I took that knowledge and eventually taught the members of my team how to hang 14 doors in a day or install kitchen cabinets in a day. Now they can do it for me and everything moves much more quickly. So my point is, figure out what you don't know and either teach yourself or learn from someone else.

ACCEPT THAT YOU NEED OTHER PEOPLE

We live in a world in which we are totally, utterly interdependent on each other, to within an inch of our lives. Most of us don't even realize it. But we depend on people to build our cars, to maintain roads, to keep the cell towers up and running—I don't need to go on. We need each other to keep our lives going the way we want them to. We don't realize no man is an island.

If you live your life or operate your business alone, thinking you're rough and tough, you are making a big mistake. Your business will stay small. Your back will hurt, you'll work long hours, you won't ever take a vacation, your wife will divorce you, and your kids will not know you. You're also going to lose money because you won't have any systems in place or people to help you when you inevitably face the ebbs and flows of the construction industry.

As soon as you acknowledge this simple fact, you will be able to accept that you need a team and will start making moves to make it happen.

SALES DONE "OUT OF HOUSE"

I suggest getting a sales guy that you pay on a commission for each closed project. Pay him as a contractor, using a 1099. The fact that you as the contractor are one person removed helps elevate you. The sales guy will go and meet the potential client and feel out the project—he'll basically qualify the prospect. When the potential becomes a hot lead then it's a great opportunity to tag you as the technical professional in the field. Not to mention, having two people on sales is better than one. Having three is better than two. If you can grow your sales force you grow your manpower and the projects will be flowing in.

A FINECRAFT PROJECT FEATURED ON HOUZZ.COM

TIPS FOR SUCCESS

List out the tasks you are involved in. Prioritize the list and isolate those things that only you can do. This separates the tasks you can hire out and it will free yourself up to train your guys or hire on more guys that you can train. The point is to free up your time so you can grow your business. Delegate anything you can hire out so you can focus on your core production team.

NOTES

CHAPTER 3

BUILDING A GREAT TEAM

NO MAN IS AN ISLAND. NO ONE CAN DO THIS ALONE. I keep saying this, but it's important. Even if you are the best contractor who's ever started a business, you can't know everything about everything. You don't know accounting and marketing and human resources. You also cannot build a house alone. So you need people who can create work that lives up to the standards you've set for your business. The bottom line—you need a great team.

Building a great team that will be committed to you and to helping you grow your business is an invaluable part of building a successful business. Unfortunately, it's not always easy. In fact, GuildQuality states that in a recent survey:

For nearly two years in a row, respondents reported that finding and hiring qualified labor is increasingly difficult. Similar to prior quarters, many are focused on hiring new sales talent to drive new leads as well as providing better training to ensure higher employee ROI.

So hiring good employees is a multi-faceted operation. In the past, people in our industry have believed that there's only one way to hire a good contractor—potential employees have to have 40 years of experience and show you a beautiful portfolio of work. Well, I've always thought that if a person with that much experience and great craftsmanship isn't busy, I had better find out why. If he's not busy because he doesn't know how to expand his own business, then you can use him and you can rely on him. But if he's not busy because he doesn't do good work or he's difficult to work with, then you will have another problem on your hands.

So sometimes, looking for the most experienced people works. Other times, it doesn't. I think there are better ways to find great people and help educate and grow them into long-term, successful and devoted employees.

At FineCraft, we have a couple of strategies we use when looking to grow our team. Our ultimate goal is to find and cultivate a pool of contractors who can take projects on and complete them in totality.

First is the more traditional route. We hold interviews and hire whomever we think is right for the job. We look at their experience and knowledge and will likely hire the person we know can take on projects on their first day. We still devote the time and attention it takes to make them as successful as possible, but we know that their experience makes them capable.

YOU CAN TEACH SOMEONE HOW TO HAMMER AND NAIL. YOU CANNOT TEACH SOMEONE HOW TO WORK HARD OR BE A GOOD TEAMMATE.

The second strategy requires a bit more time and attention, but, in the end, has resulted in devoted, knowledgeable team members who have helped propel FineCraft into greater success.

Not every person who would be a right culture fit with your company has all of the necessary knowledge and experience it takes to be a successful contractor. But it's important to give as much weight to the characteristics that would make them valuable—intelligence, a willingness to learn, etc.—as you do their experience. After all, you can teach someone how to hammer and nail. You cannot teach someone how to work hard or be a good teammate.

As a result, sometimes we hire people who have very little experience. They might not be able to adequately read plans or build a house, but I have the utmost confidence that I can teach them everything I know, and, based on their personality and characteristics, they'll end up being the best possible employee for my business.

We do this strategically, as ultimately we are looking for leaders who can manage a project from start to finish. Many contractors use subcontractors for every project, parceling out the roof and the electric and the painting. We do that, too. But we're not really relying on that as the future of our business. We're looking to find people who can serve, essentially, as general contractors for each project. We train them to take the project and do the whole thing. Then we take a percentage of the earnings. It works to a very large extent, keeping FineCraft profitable, but takes a lot of the stress off of me, in terms of managing every project that comes through our doors.

So with that strategy in mind, I'm going to explain how we determine who will make a great leader, even if they're lacking in experience.

I BELIEVE THERE ARE SIX MAIN CHARACTERISTICS to look for when hiring team members:

- **Love for the Job**
- **Willingness to Learn**
- **Intelligence**
- **Able to Deal With Adversity**
- **Work Ethic**
- **Honesty/Integrity**

I'm going to expand on each of these qualities and explain why they're important in this hiring process.

• LOVE FOR THE JOB

The most important thing is that a potential employee loves what he's doing. This job isn't always easy. There will be unhappy clients, crappy weather, difficult projects, and more. If he doesn't love the job, he won't last long in the face of these challenges.

• WILLINGNESS TO LEARN

There is always something to learn in the contracting business. There are always new trends, new techniques, new materials, and new client demands. Even after all these years, I am still learning. I never want my employees to think they know everything. If he's not willing to learn, we will know immediately and he won't be able to stay. If someone is willing to learn, I can take a responsible and respectful employee and teach them everything they need to know to successfully take on a project and finish it.

There's an old story us salty old timers tell, about the new employee who just got hired. The carpenter gives him the keys to the truck and tells him, "Go get me the woodstretcher." The new employee has no idea what a woodstretcher is, but he runs out to the truck, where he fumbles around, spending valuable time looking for something he can't even identify. Eventually, he has to go back to the contractor and ask. He should have just asked the question right away instead of wasting time.

I have found invariably that new contractors come in bright-eyed and bushy tailed, just saying, "Yes, yes, yes, I can handle anything!" Now, don't get me wrong. I appreciate confidence and a willingness to work hard. After all, there were plenty of times when I first started my business, I told clients I could do everything.

But don't be fooled. They'll make you think they're really smart and willing and able. And they might be. But part of that is a willingness to ask questions and always being ready to learn. Which leads to the next quality I look for in a contractor.

• INTELLIGENCE

Most people think it doesn't take a lot of intelligence to be a contractor. And if you're only talking about swinging a hammer or rolling paint onto a wall, you might be right. But being a true contractor is a different story. It takes literal intelligence, so you can read plans, order supplies, etc., but also the emotional intelligence to deal with clients and manage a team.

If someone is intelligent, I can teach them everything they need to know to be successful. I just need someone who is bright enough to follow plans and instructions. I have worked out some systems very specific to the contracting business that help my employees easily read plans and order supplies. I will expand on those in later chapters.

But I do want to caution against writing off new employees too quickly. These bright-eyed, bushy-tailed guys have to go through a period of stupidity. I mean literal stupidity. You just have to push them through this time period and bring them out the other side.

If they are intelligent enough, they will let you lead them through it, and learn to ask questions and learn from those of us with more experience. If you can lead them through it, they will be able to comfortably face, confront, and deal with adversity and gain invaluable experience.

• ABLE TO DEAL WITH ADVERSITY

I want my employees to be able to effectively deal with confrontation. I don't necessarily mean a fight or an argument. I mean he has to be able to face and handle adversity comfortably. The key word is comfortably.

There are always going to be problems to confront in this business. If there is a leak in the plumbing, you have to confront where it's coming from, instead of running around and saying "We've got a leak in the whole works!"

It isn't going to stop raining because you forgot to put plastic over an addition. It's gonna fucking rain. Someone is going to mess up a supply order and you're not going to have enough tile. You have to be able to calmly address the mistake and order the fucking tile. If you have a problem, you have to find out what the problem is and fix it.

If you're not able to confront a problem, determine its source, and fix it without panicking, you're going to get screwed.

• WORK ETHIC

Contracting is hard work. It's physically hard, it's demanding on your time, with early days and late nights. Employees can't get away with skating by, doing just the minimum, because it will become clear that nothing is happening and moving forward on the job site.

I need people who are always willing to work hard. Even if they aren't always perfect or can't handle every aspect of a job, I want them to jump in. I'm looking for the people who, if they don't have a specific piece of the project to finish, are cleaning up, gathering trash, and putting away tools. I'm looking for the people who aren't afraid to tell me that they think this part of the project needs to be corrected or looked at more closely. I'm looking for people whose work ethic stands out in everything they do.

• HONESTY/INTEGRITY

I have no patience for people who aren't honest with me. Nothing good comes out of not telling the truth. I need employees who are going to be honest with me about whether or not they can handle a job or when they think they can complete a project. I need employees who are willing to tell the truth to homeowners, even when it's not what they want to hear.

A FINECRAFT PROJECT FEATURED IN HOME & DESIGN

NOTE FROM THE EDITOR:

George does not mince words when it comes to just what kind of team he wants. He does not suffer fools and doesn't have time for people who "aren't using their brains." It's a refreshing honesty that I'm sure can be a bit abrasive to some, but one that seems to serve him well in business.

In turn, his employees seem to respect him, his opinion, and his experience. They are confident in their work but appreciate George's input. George speaks to them as colleagues and respects their experience also.

At the company Christmas party, one of George's long time colleagues told me, "George does things right the first time. He's building a team of honest and smart guys who can handle the business."

NOTE!

SO WHAT DO THESE SEVEN CHARACTERISTICS LOOK LIKE IN PRACTICE?

I have one employee in particular who checks all of the boxes of these seven characteristics.

He's not perfect. He sometimes has difficulty prioritizing his workload and confronting adversity. He ends up running around like a chicken with its head cut off when things go wrong.

But he's an integral part of what we do. He fills all the odds and ends of what needs to happen. He is always working hard. He's honest with me and he's intelligent enough to juggle a very heavy workload.

But he cannot nail one nail. If you give him a nail, he would try to nail it with his thumb. So he knows nothing about building a house, but our business wouldn't be the same without him. We worked hard to identify his strengths and then develop his talents into the best job for him—one that also helps our business.

MAKE PLANS EASY TO UNDERSTAND

You can't expect that every contractor you hire has enough of a background in architecture or construction that they can immediately read a set of plans. Many of them will need help in order to learn how to accurately and effectively read plans and build from them. Taking the time to help them learn will help your business immeasurably.

I learned early on that if I wanted my employees to be able to read from and work off of plans, I had to ensure that they could read them. Simple as that. So I put a lot of time and effort into developing a simple and effective system for sharing plans with my crews.

CONSTRUCTION SHEETS

The first thing I do on a project is what I call "construction sheets." The construction sheets are simply a repetition of what I did in my estimate (which we will talk about in a later chapter). I lay out every single thing we know every project has to have. For example, every project has footings, has foundation walls, has floors, has walls. I start with a very quick sketch of each one of these elements. Then I premeasure from the architect's measurements, and I know exactly how much wood needs to be cut, how much flooring we need, etc.

So, for example, if we have a house, or an addition, you would have like three or four walls, or maybe eight if there's a unique shape. But it would be very easy from the plans to get the perimeter dimensions. So I create a sketch that has those perimeter dimensions, and once we start excavation or whatever type of construction, we respect those points that were written down on a piece of paper. I have premeasured from measurements, and I have exactly how much each one of these pieces of wood would be cut. That way nobody has to guess.

We start with the floor. A floor consists of joists, and those joists and supporting elements, rest on a wall. Then on top of that you put plywood. And you've created the sub-floor. On top of the sub-floor you put other finishes. But once you start with the floor, everything else is dependent on that sub-floor. We then build out each part of the project individually.

IT GOES FROM A PLAN THAT'S HARD TO READ TO A SIMPLE DIAGRAM...

At the end, we're not looking at the architect's plans anymore, but my construction sheets. Why? Because the architect's plans, although they're handy to use at any time, have too much other information. There are elevations, measurements, and so much more. There's so much information that unless you really know how to read plans, it's very easy to be confused.

I take all the extra stuff the architect, in his infinite wisdom, has shown on one page, and I distill it into the things that we need right this moment for our purposes. It goes from a plan that's hard to read to a simple diagram that can be digested by even the most inexperienced individual.

It's very simple, but the simplicity of it should not be confused with its crucial importance. Because it's so important I've made it so simple. So people cannot fuck it up.

So when my crews start they don't have to think too much. They'll know they have to cut 268 inches and a half, 325 inches and three-quarters, and 99 and five-eights. And they cut those, and they put them where they're supposed to go, and miraculously a floor appears.

Then I do the same thing with walls. Now that we have the floor done, we have to build walls. The walls are built out of vertical pieces and horizontal pieces. Each wall, no matter how many walls there are, I have them all pre-measured and pre-marked. All of them are cut within an eighth of an inch, including the window openings. The windows magically created out of nowhere, and suddenly there's a window there, and a door here, because the measurements were all done. The pieces come together like a jigsaw puzzle.

ON THE NEXT FEW PAGES, YOU'LL FIND SOME SAMPLE PAGES OF MY CONSTRUCTION SHEETS.

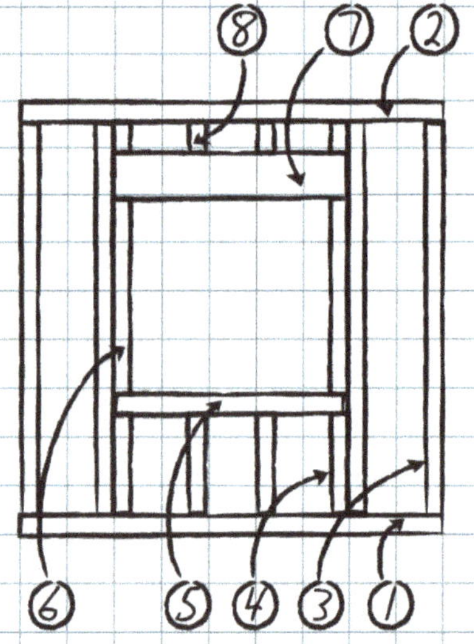
8
7
2
6
5
4
3
1

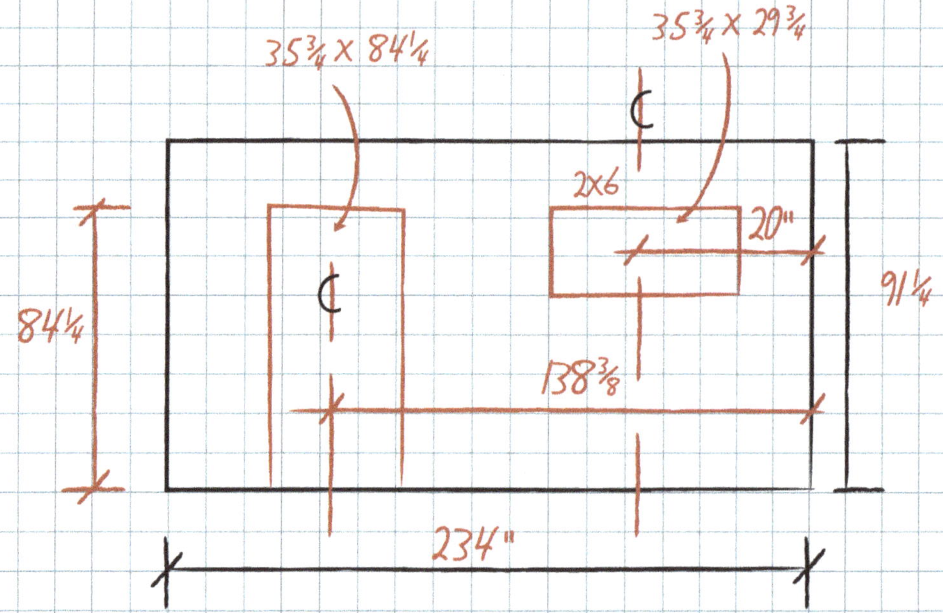
35¾ X 84¼
35¾ X 29¾
2X6
20"
84¼
91¼
138⅜
234"

WALL # 2

VIEWED FROM INSIDE.

PROJECT:	JOHN DOE
FLOOR #	INT (EXT)

#	TYPE	SIZE	QTY	REMARKS
1	BOT. PLATE	234'	1	
2	TOP PLATE	234'	1	
3	STUDS	86¼	14	
4	JACKS-UNDER	54½	6	
5	WIND SILL	41¾	1	
6	WIND JACK	29¾	4	
7	(3-2X6) WIND HEADER	41¾	3	SANDWICH BETWEEN THEM CUT ONE TOP PLATE
8	JACKS-OVER	—		
9	DOOR JACKS	41¾	4	CUT ONE TOP PLATE
10	(3-2X6) DOOR HEADER		3	

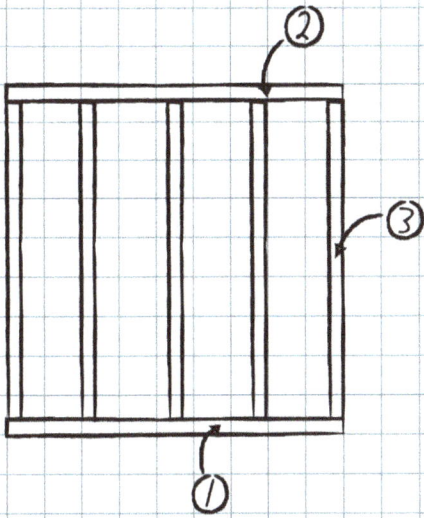

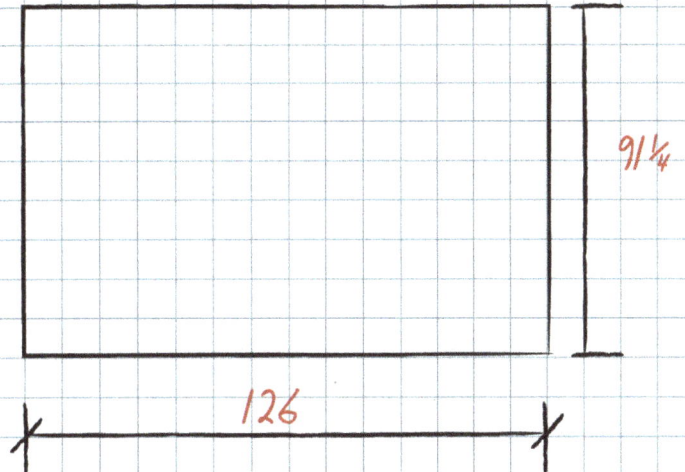

PROJECT:	JOHN DOE
FLOOR#	INT (EXT)

WALL # ①   VIEWED FROM INSIDE.

#	TYPE	SIZE	QTY	REMARKS
1	BOT. PLATE	126"	1	
2	TOP PLATE	126"	1	
3	STUDS	86¼	10	

WHEN YOU'RE LOOKING AT PLANS, you're constantly going back and forth between elevations, different sections, etc. So the other thing I do is highlight each part of a plan. I have five or six different colored highlighters, and I color-code everything. For example, interior walls will be marked one color, exterior walls would be marked another color. Pressure treated wood would be marked another color. Regular wood is marked another color. And on and on.

I highlight all the walls from an elevation or floor plan, so they pop out and it becomes very clear where things are. It sort of makes all the other lines and everything else disappear. You can really start seeing very clearly what you have in front of you, and you don't have to be affected by all the other junk in the tremendous amount of data that is crammed into a floor plan or an elevation. This also helps in my estimating process, which we'll cover in the next chapter.

I've found that this speeds up our process in amazing ways. Our crews don't have to spend a ton of time looking at the plans, reducing the chance of them making mistakes. They know all the dimensions before they even start cutting anything.

I'VE FOUND THAT THI. SPEEDS UP OUR PROCES. IN AMAZING WAYS.

We have all the answers in the blueprint, but I've just distilled it into something that's just easier to understand. At some point you have to give them the information anyway, and if giving it to them packaged this way is better, easier, and faster for everyone, then it 100% makes sense. I present the information in a different and better way for their purposes. So they can do it faster and more accurately.

Basically I've tried to standardize, pre-measure, and pre-cut everything, so my crews can just put it together. The bright ones who pick it up move like startled gazelles. They can move quickly and efficiently and become an asset to your team, getting work done in a fraction of the time.

TIPS FOR SUCCESS

We've included a blank, printable Construction Sheet for you so you can begin to use them in your own business today!

WALL # ◯   VIEWED FROM INSIDE.

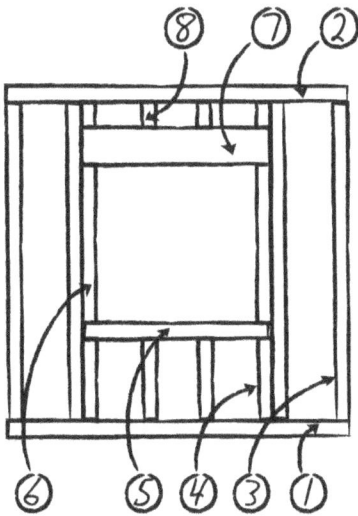

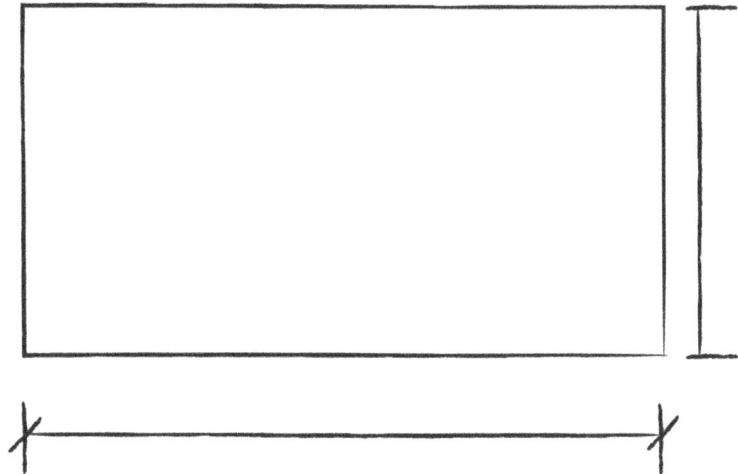

PROJECT:	
FLOOR #	INT EXT

#	TYPE	SIZE	QTY	REMARKS
1	BOT. PLATE			
2	TOP PLATE			
3	STUDS			
4	JACKS-UNDER			
5	WIND SILL			
6	WIND JACK			
7	WIND HEADER			SANDWICH BETWEEN THEM
8	JACKS-OVER			

② ③ ①

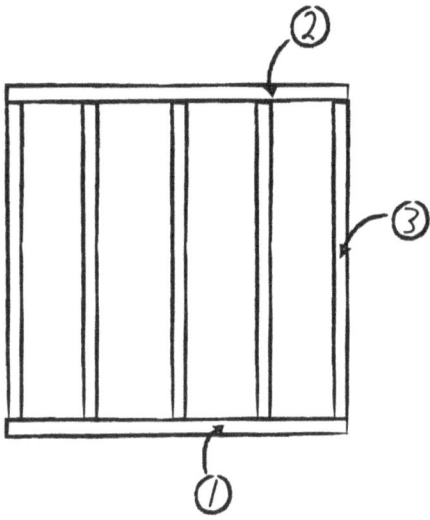

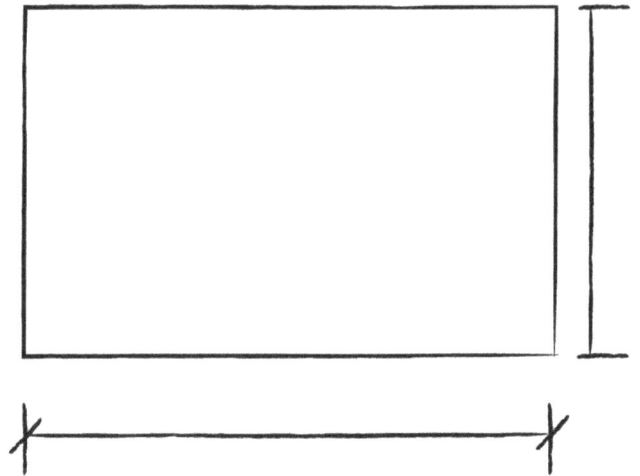

PROJECT:	
FLOOR #	INT EXT

WALL # ◯   VIEWED FROM INSIDE.

#	TYPE	SIZE	QTY	REMARKS
1	BOT. PLATE			
2	TOP PLATE			
3	STUDS			

NOTES

CHAPTER 4
COMMUNICATION

I DON'T THINK THAT ANY GOOD RELATIONSHIP, whether it's a working relationship or a personal one, can survive or thrive unless there's good communication. Communication is especially important in the construction business, because as we all know, things can turn on a dime and you need to have systems in place that will help you, your crew, and your customers make decisions and address problems as they arise.

COMMUNICATION WITH CUSTOMERS

At FineCraft, we make communication with our customers a top priority. We start by making sure that not only are we a good fit for a customer's project, but also that the customer is a good fit for us. Before we initiate a first meeting with a prospect we send them a questionnaire to determine, basically, what kind of a client they would be. Because as much as we want to keep our business growing, we're not willing to sacrifice our sanity to deal with clients who are going to cause more problems than they're worth or who would simply be better off with a different contractor.

We recently had a client complete a questionnaire and we knew right away that it wouldn't be a good fit. He wasn't willing to spend the money we would charge for his project and it would have caused problems in the end. He would have been unhappy, as would we. So we didn't take the project. It might feel hard to pass up money, but it's not always worth it.

After we have decided to work with a client and they've chosen us, we put a regular communication schedule in place. We start by putting their information into our email system so they start getting regular communications from us and begin to learn about our history.

OUR CLIENTS ARE SENT A DETAILED EXPLANATION OF WHAT THE PROCESS WILL LOOK LIKE. ON THE NEXT PAGE IS A SAMPLE OF OUR SALES FUNNEL AND PROCESS ON ANY PROJECT WE TAKE ON.

FINECRAFT SALES PROCESS

STEP 1: INTRODUCING OURSELVES.

We'll meet with you and provide you with a complimentary site visit and consultation. If you have questions we'll provide answers. If you have plans ready to be estimated we'll provide you with an estimate. If you don't have plans and need assistance we'll provide it. Need advice on where to start with your research of architects? Getting on the right track can be daunting, so send us an email with your questions. Click here to start a conversation.

STEP 2: GETTING TO KNOW THE FINECRAFT FAMILY.

The next step is getting to know each other. You know the expression "like an open book?" Well, that's us. FineCraft is an open book in the sense that we want you to get familiar with us, with our work, with our workers, and with our past clients. Therefore, the next step to take while waiting for your estimate is a tour of our projects. Come see our work for yourself and get a feel for what it's like to have a project done by FineCraft. Click here to schedule your tour.

STEP 3: ESTIMATING YOUR PROJECT

In working up the numbers for your project a walkthrough and a conversation might be enough, but most of the time we'll need plans to properly and more accurately bid your project. And we like to walk the project once, maybe twice, to get a better idea about the details and scope of the project. Estimating a project depends on size and workload. Since we work on estimates in the order in which we get them, yours may be third in the queue. Therefore, it's not unusual to wait 2-3 weeks before getting an estimate back. We'll keep you informed about our progress. Have plans to send us? Click here to send us your plans today.

STEP 4: GOING OVER THE ESTIMATE

Once we deliver the estimate you may need assistance in understanding the details, and that's why we're here. We consider it our job to help inform you about the nuances of your project. If you need help prioritizing, maybe even phasing out the project over time, we can help you with that, too. Estimating is one thing. Understanding it is a whole new territory. We'll go over your "allowances" and "scheduling." We'll even help you by suggesting vendors for different aspects of your house, i.e. who to go to for what, and even what it means to get a "Contractor's Discount."

STEP 5: GETTING STARTED

Before we get started on your project we have a Preconstruction Meeting to go over the starting date, hours of operation, what door to use to get in and out of the house, what plants to save, etc. At this point if we haven't gone over the Important Terms we then go over them along with whom to contact for what. Click here to download our Client Information and Common Questions PDF's.

STEP 6: YOUR TEAM

You'll have a team that will be in charge of the day-to-day tasks as well as ordering and receiving materials, executing the job and ensuring all goes according to plan. This includes coordinating inspections along the way. If you have any questions you'll always have a Project Manager and a Project Coordinator to go to.

STEP 7: FINISHING THE PROJECT

The work we perform is guaranteed for one year. Within this timeframe, should anything that occurs that pertains to our work, you should feel free to contact us so that we can address what's needed. You'll have a designated person to address things that need a little TLC after the project is completed.

Finally, we'll look at the calendar, determine when the project will be complete and break it into quarters. Then we break it down further to schedule regular meetings with clients. We have regular meeting times with homeowners, on site, every two weeks. Whenever I can be at those meetings, I am. Customers appreciate that the owner of the company takes time to come to their home, check in, and address any problems.

Throughout the project, we survey the client along the way. It helps us know whether they're happy with how the project is going, what they'd like to do differently, etc. And then of course, we'll survey the client when we're done with the project so we can learn and do better next time. You'll find a sample of our customer questionnaire at the end of this chapter.

COMMUNICATION WITH OUR TEAM

When it comes to communication and scheduling within our business, it's also a priority that we're all aware of status and on the same page for each part of the project. I check in with my crew leaders every single morning and throughout the day.

We have a checklist with milestones for each part of a project from ordering to construction. We set daily and weekly targets for each part of the project. We know each date that each phase of the project should be complete by. All of our crews know these dates and what's expected from them in order to meet those goals.

WE'VE INCLUDED A PRODUCTION CHECKLIST AT THE END OF THIS CHAPTER FOR YOU!

Ⓝ

NOTE FROM THE EDITOR:

This is one of the first things I noticed about George: He is always in constant contact with his team. He's on the phone constantly, talking to his crew leaders, suppliers, anyone who needs his attention. He acknowledges texts as soon as he possibly can. He told me, "Management equals caring about the customer and my employees, and keeping that communication open. Without that, any other system or method just won't work. If you care, you'll find a way to make anything work."

George also remembers everything, whether it's how much money they owe the tile company to what inspections are due when. When I asked him how he does it, he laughed and said he didn't know.

NOTE!

ACCURATE ESTIMATES

One of the biggest ways we keep our timelines accurate and communication open and easy starts with our estimates. I have a very detailed and simple way that I complete estimates. It ensures accuracy every time.

I have drilled down a list of everything a project has to have, whether it's a complete new build or a remodeling project. Everything from nails to painter's tape to wood to insulation. Then, as I'm estimating the cost of the project, I use the worksheets I've developed, along with the color-coded floor plans we use for construction. Each piece of the project is drilled down into smaller pieces so we don't miss anything and there aren't any surprises later.

THE NEXT FEW PAGES CONTAIN SEVERAL SAMPLES OF MY ESTIMATE WORKSHEETS SO YOU CAN SEE THEM IN ACTION!

A FINECRAFT PROJECT FEATURED IN HOME & DESIGN

TIPS FOR SUCCESS

We've included a blank Customer Survey, Production Checklist, Punchlist, and Estimate Worksheets here for you to begin using today!

ESTIMATE SHEET

JOB NAME: ADDRESS:

ITEM	PRICE/ UNIT	PICTORIAL FORM	QTY	PRICE
EXCAVATION, FOOTINGS, FOUND WALL, WATERPROOFING, DRAINAGE				
EXCAVATION, BACK- FILL, GRADING FOOTINGS	22.09/LF			
REINFORCEMENT	2.18/LF			
FOUNDATION WALL 8" THICK 36" HIGH	14.04/LF			
PARGING	4.0/LF			
WATERPROOFING	2.9/LF			
DRAINAGE	5.4/LF			
TOTAL	$51.42			
CONCRETE SLAB	/SF			
4" CONCRETE SLAB w/ 6x6 WIRE MESH & 6 MIL VAPOR BARRIER	$4.12/SF			
4" GRAVEL	.73/SF			
TOTAL	$4.85 /FT			
EXTERIOR WALL BRICK VENEER 10'-0" HIGH x 1'-0" WIDE WALL				
BRICK	$102.80			
2X4 FRAMING	$1.17			
1/2 SHEATHING	$1.52			
R-11 FOIL-FACE INSUL	$7.00			
1/2" DRYWALL	$1.80			
3-PIECE TRIM BASE	$1.00			
INTERIOR PAINTING (INCLUDES PREPARATION) PRIME & 2 COATS				
TOTAL	$159.59/FT			
				TOTAL

ESTIMATE SHEET

ITEM	PRICE/UNIT	PICTORIAL FORM	QTY	PRICE
EXTERIOR WALL	WOOD SIDING	10'-0" HIGH X 1'-0" WIDE		
2x6 FRAMING	$15.30/FT			
½" SHEATHING	10.00 "			
REDWOOD SIDING 8"	32.10 "	STUDS		
R-19 INSULATION	9.20 "	1ST FLOOR		
½" DRYWALL	13.36 "	2ND FLOOR		
3-PIECE BASE	5.33 "	ATTIC		
EXT PAINTING	10.01 "			
INT PAINTING	8.40 "			
(INCL PREP, PRIME & 2 COATS)		10'-0"		
		12"		
TOTAL	$103.70/FT		67	8,100
INTERIOR WALL	$/FT	10'-0" HIGH X 1'-0" WIDE		
2x4 FRAMING	$12.60/FT			
½" DRYWALL 2 SIDES	26.72	STUDS	67	6,700
3-PIECE TRIM 2 SIDES	10.66			
PAINTING 2 SIDES	16.80			
(INCL PREP, PRIME & 2 COATS)		10'-0"		
		12"		
TOTAL	$66.78/FT			
FIRST FLOOR	$/FT			
2x10 JOISTS 16' OC	$3.90			
R-30 INSULATION	$1.17			
¾ PLYWOOD	$1.52			
¾ OAK FLOOR FINISHED	$7.00	FLOOR JOIST		
½" PLYWOOD UNDER	$1.80			
PAINTING OF PLYWOOD	$1.00			
		12" 12"		
TOTAL	$15.89/FT		435	8,700
			TOTAL	23,500

ESTIMATE SHEET

JOB NAME: ADDRESS:

ITEM	PRICE/ UNIT	PICTORIAL FORM	QTY	PRICE
SECOND FLOOR	/SF			
2X10 JOISTS 16" OC	$3.90			
3/4" PLYWOOD	1.62			
3/4" OAK FLOOR (FINISHED)	7.00			
1/2" DRYWALL	1.40			
PAINTING OF DRYWALL	.54			
TOTAL	$14.46			
ROOF CEILING	/SF			
2X6 JOISTS 16" O.C.	2.22			
R-30 INSULATION	1.17			
1/2" DRYWALL	1.40			
PAINTING OF DRYWALL	.54			
TOTAL	$5.33/SF			
ROOF	$/SF	10'-0" HIGH X 1'-0" WIDE WALL		
2X8 14" O.C.	$5.91			
1/2" SHEATHING	$1.20			
ASPHALT SHINGLES	$2.00			
TOTAL	$9.11 /SF			
			TOTAL	

FLOOR JOIST   12"   12"

FLOOR JOIST   12"   12"

RAFTER   12"   12"

ESTIMATE SHEET

ITEM	PRICE/ UNIT	PICTORIAL FORM	QTY	PRICE
END ROOF	/LF			
		JUDGEMENT CALL EXT TRIM →		
FASCIA BOARD 1x8	$2.61			
SOFFIT 8"	3.03			
GUTTERS/DOWNSPOUTS	4.47			
PAINT OF WOOD 2 COATS	2.79			
2" CONT. VENT	3.35			
		12"		
TOTAL	$16.25/LF			
DOORS				
		CASING EXT. BAS. EXT. 1ST EXT 2ND BAS INT 1ST 2ND		
COST OF DOOR	VARIES			
INSTALLATION	$60.00			
JAMBS, CASING BOTH SIDES	113.00			
HINGES, HARDWARE	20.00			
PAINTING BOTH SIDES - 2 COATS	89.00			
TOTAL	$282/DOOR + COST OF DOOR ≈600			
WINDOWS		10'-0" HIGH x 1'-0" WIDE WALL		
		BAS. 1ST 2ND		
COST OF WINDOW	VARIES			
INSTALLATION	$50.00			
INTERIOR TRIM	50.00			
EXTERIOR TRIM	50.00			
INT PAINTING 16 LIGHTS	53.00			
EXT PAINTING 16 LIGHTS	53.00			
TOTAL	$266/WINDOW + COST OF WINDOW ≈1200			
				TOTAL

ESTIMATE SHEET

DATE: PHONE:

JOB NAME: ADDRESS:

ITEM	PRICE/ UNIT	PICTORIAL FORM	QTY	PRICE
DEMOLITION				
TOTAL				
CONCRETE				
CONCRETE STEPS				
BRICK & 4" BLOCK WALL	12.69/SF			
BRICK PIERS 16" SQ.	37.22/LF			
BRICK FIREPLACE				
BRICK CHIMNEY				
BRICK PATCH				
BRICK PATIO /SF				
TOTAL				
STEEL BEAMS				
WOOD BEAMS				
FLITCH PLATES 3/8"x9"	$21.00/LF			
DIFFICULT FRAMING				
TOTAL				

TOTAL

DATE: PHONE:

ESTIMATE SHEET

JOB NAME: ADDRESS:

ITEM	PRICE/ UNIT	PICTORIAL FORM	QTY	PRICE
DECK				
SURFACE				
FRAMING				
WOOD POSTS	$10.00/SF			
FOOTINGS				
PICKETT RAILING	$17.61/LF			
STEPS	$39.68 EA			
SEAT	$7.00/LF			
TOTAL				
PLUMBING				
TOILET				
LAVATORY				
BATH TUB				
KITCHEN SINK				
WASHING MACHINE				
GAS DRYER				
GAS RANGE				
ICE MAKER				
HOT WATER HEATER				
GROUND WORK				
TOTAL				
HVAC				
CAST IRON RADIATORS				
BASEBOARD WATER/ ELECTRIC	$21.00/LF			
BOILER/FURNACE				
ELECTRIC FURNACE				
HEAT PUMP				
GAS FURNACE				
A/C				
DUCTWORK				
HUMIDIFIER				
TOTAL				
				TOTAL

DATE: PHONE:

ESTIMATE SHEET

JOB NAME: ADDRESS:

ITEM	PRICE/UNIT	PICTORIAL FORM	QTY	PRICE
ELECTRICAL				
DUPLEX OUTLETS	49.5 EA			
APPLIANCE OUTLETS	90. EA			
220-V OUTLET	200. EA			
SWITCHES	46.5 EA			
CEILING FIXTURES	60. EA ← NO FIXTURE			
BATH FIXTURES	60. EA ← NO FIXTURE			
RECESSED FIXTURES	130.00 EA ← INCL. FIXTURE			
SPOTLIGHTS	250.00 EA ← INCL. FIXTURE			
DIMMER SWITCHES				
TOTAL				
INCREASE SERVICE	900-			
BATHROOM FANS	225.00			
KITCHEN FAN	320-			
RE-LOCATE SERVICE	500.00			
TOTAL				
PLASTER PATCHING				
WOOD PANELLING				
TILES				
CERAMIC	10.88			
QUARRY				
MARBLE				
VINYL				
OAK FLOOR INSTALL & FINISH	$7.00/SF			
↳ REMODEL				
TOTAL				
				TOTAL

DATE: PHONE:

ESTIMATE SHEET

JOB NAME: ADDRESS:

ITEM	PRICE/ UNIT	PICTORIAL FORM	QTY	PRICE
BUILT-IN BOOKCASES	12.78/SF	SF=FRONT SQUARE FOOTAGE #2 WOOD		
SHELVES				
LINEN CLOSET	$66.00/SF			
CLOSET	18.00/SF			
ATTIC STAIRS				
MAIN STAIRS OAK	2500			
WROUGHT IRON				
OAK BOOKCASES	20.00/SF	SF=FRONT SQUARE FOOTAGE		
TOTAL				
KITCHEN CABINETS				
COUNTERTOPS				
SINK				
DISPOSAL				
DISHWASHER				
RANGE & OVEN				
HOOD				
HOOD VENT				
SHOWER DOORS				
MEDICINE CABINETS/MIRRORS				
TOTAL				
METAL FIREPLACE	1100			
ATTIC EXHAUST FAN				
GARAGE DOORS				
GARAGE DOOR OPENERS				
ALARM SYSTEM				
SMOKE DETECTORS				
SKYLIGHTS				
STORM DOORS				
FENCE				
TOTAL				

TOTAL

ESTIMATE SHEET

JOB NAME: ADDRESS:

ITEM	PRICE/ UNIT	PICTORIAL FORM	QTY	PRICE
MISC PAINTING WALLPAPERING				
TOTAL				
TRASH PICK-UP CLEAN-UP				
TOTAL				
TOTAL				

TOTAL

DATE:

PHONE:

ESTIMATE SHEET

JOB NAME:

ADDRESS:

ITEM	PRICE/ UNIT	PICTORIAL FORM	QTY	PRICE
TOTAL				
TOTAL				
TOTAL				

PROJECT ESTIMATE

DATE:_____

BILL TO: JOB NAME:

_____   _____

_____   ❑ Work Completed   ❑ Estimate

_____   Date Completed_____

DESCRIPTION	AMOUNT
TOTAL	

SIGNATURE (OWNER)   SIGNATURE (CONTRACTOR)

PROJECT ESTIMATE

DESCRIPTION	AMOUNT
TOTAL FROM FIRST PAGE	
TOTAL THIS PAGE	
GRAND TOTAL	

JOB ESTIMATE

DATE:_____

BILL TO: JOB NAME:

_____   _____

_____   ❑ Work Completed   ❑ Estimate

_____   Date Completed_____

Itemized Breakdown	Amount
Excavation, footings, foundation wall, backfill, grade	
Concrete slabs (per plans & specifications)	
Exterior walls (brick, frame, 1/2" plywood, insulation, 1/2" drywall, trim, paint)	
Exterior walls (siding, frame, 1/2" plywood, insulation, 1/2" drywall, trim, paint)	
Interior walls (frame, 1/2" drywall, trim, paint)	
First floor (frame, insulation, 3/4 plywood, oak, finish)	
Second floor (frame, 3/4" plywood, oak, finish) (1/2" drywall, paint)	
Ceilings (frame, insulation, 1/2" drywall, paint	
Roof (frame, 1/2" plywood, tar paper, shingles-slate-membrane, metal)	
End-Roof (fascias-soffits, gutters, downspouts) (copper-aluminum)	
Exterior trim (per plans)	
Exterior doors (purchase, installation, trim, paint, hardware)	
Interior doors (purchase, installation, trim, paint, hardware)	
Windows (purchase, installation, trim, paint, hardware, screens)	
Demolition (per plans)	
Concrete	
Concrete steps	
Brick & block wall	
Brick piers	
Brick fireplace/chimney	
Brick patch	
Patio (brick-stone) on (concrete-blue stone)	
Beams (steel-wood)	
Misc. Framing (custom, soffits)	
Sub-total	

JOB ESTIMATE

ITEMIZED BREAKDOWN	AMOUNT
Deck (footings, posts, 5/4 - 2x4 - 2x6 decking, stairs, railing)	
Pergola	
Plumbing (including fixtures, gas line, hot water heater)	
Heating (radiators-new furnace, base board radiators	
HVAC (forced air system) (A/C, furnace, heat pump, extend exist)	
Electrical (including fixtures, heavy-up)	
Plaster-Drywall patching (connect old to new)	
Tiles (marble, ceramic, quarry, vinyl, labor-mat incl.)	
(allowance for purchase $ /sf) (per plans)	
Oak floors (refinish existing-new-patch)	
Built-ins - Bookcases (allowance/per plans)	
Misc. Int. trim work, shelving	
Stairs (main, railings, attic)	
Wrought iron work	
Kitchen cabinets (installation only/purchase allowance $	
Kitchen counter tops (laminate-Corian-Granite) (installation/allowance $)	
Shower enclosures	
Mirrors-shower rod-medicine cabinets-accesories (bath)	
Metal fireplace (purchase-installation-hearth-mantel-marble-granite)	
Skylights (purchase, installation, finish)	
Storm/screen doors	
Garage door (w/openers)	
Misc. painting interior	
Misc. painting exterior	
Trash pick-up/clean up	
Portable toilet/telephone hook-up	
Supervision-Administration of project - Profit & Overhead	

ESTIMATED JOB TOTAL

ESTIMATED BY

This estimate is for completing the job as described above, it is based on our evaluation and does not include material price increases or additional labor and materials which may be required should unforeseen problems or adverse weather conditions arise after the work has started.

ALLOWANCES

In construction, an allowance is an amount specified and included in the construction contract (or specifications) for a certain item of work (e.g., appliances, lighting, etc.) whose details are not yet determined at the time of contracting.

INCLUDED	EXCLUDED	ITEM
☐	☐	Tiles
☐	☐	Plumbing Fixtures
☐	☐	Elecrical Fixtures
☐	☐	Cabinets
☐	☐	Bath Vanities
☐	☐	Tops Kitchen & Bath
☐	☐	Appliances
☐	☐	Mirrors
☐	☐	Bath Accessories
☐	☐	Shower Doors

Others not specified above:

INCLUDED	EXCLUDED	ITEM
☐	☐	
☐	☐	
☐	☐	
☐	☐	
☐	☐	
☐	☐	
☐	☐	
☐	☐	
☐	☐	
☐	☐	
☐	☐	

PUNCH LIST

DATE_____

Name:_____

Address:_____   Home Phone:_____

City: _____   Office Phone:_____

State: _____ Zip:_____   Email:_____

PUNCH LIST INFORMATION

This form is used after Substantial Completion to both facilitate and ensure the completion of your project 100%.

Substantial Completion is when the the project is finished to the point where the home owner can occupy the space. At this stage you are accepting the project as-is and are ready to live in the space under the intent it was constructed. (By this time Substantial Completion should be done.)

As mentioned, this Punch List is used after Substantial Completion, which means we will address any issue that is written down on this form. For example, there might be a small patch of paint on a window in the second floor guest room, or an electrical plate in the entryway of the house which needs to be replaced. These sort of items should be noted by you.

At this point you (the home owner), the architect and the representative will do a walk-through of your project writing down what items need to be addressed. These points need to be written down on this Punch List Form provided. Once the Punch List is complete there will be another walk-through with you, the architect and the   representative to ensure its 100% completion and your satisfaction. Please approve each item that is addressed by your initials after you see that they are in fact done.

After the items in the Punch List have been addressed you, the architect and the   representative will sign this form establishing that the Punch List stage is 100% complete without further items needing attention.

1-Year Warranty Information

Issues that need to be addressed after this will be taken cared for under the 1-Year Warranty which started after the completion of Substantial Completion: _____, and which will end on _____.

It should be noted that any accidental damage or abuses which occurred after the completion of this Punch List on _____ shall not be covered under our 1-Year Warranty. *(Please see our 1-Year Warranty Terms and Agreements for more details.)*

ITEM NO.	LOCATION	DESCRIPTION	APPROVAL

PUNCH LIST

Item No.	Location	Description	Approval

PUNCH LIST

ITEM NO.	LOCATION	DESCRIPTION	APPROVAL

On this day _____ of _____, 20_____, we hereby agree that the items in the Punch List Form are 100% complete and don't need further attention. Issues that arise at the close of this Punch List will be addressed as Warranty Items.

Home Owner: _____

Architect: _____   Representative: _____

Customer Completion Survey

OVERALL SCORE:

Date:

Project Description:

Client Name:

Project Foreman:

Referral:

1.   How do you think the project was run?

2.   In your opinion what can we be better at doing?

3.   Please rate the following:

(Ratings: **1** = Poor **2** = Fair **3** = Good **4** = Very Good **5** = Excellent)

- **Friendliness of staff** –
- **Efficiency** –
- **Cleanliness** –
- **Attention to detail** –
- **Appearance** –
- **Management of your account** –
- **Communication** –
- **Follow up on problem resolution** –

4. Do you have anything else you'd like to say about how the project was run?

5. What was the deciding factor to go with us as your contractor?

6. Are you planning on doing any other upgrades to your house in the near future?

7. If so, when?

8. If money is an issue then offer EnerBank. (If not go to #10)

9. Offer to send EnerBank link.

10. Is there anyone that you know that is interested in doing anything to their house?

11. Anything you recommend we do to help you with these prospects?

File this in a file labeled **"Customer Completion Surveys"**

12. Would you write a letter of recommendation on us?

13. Houzz review?

14. Can we add you to our list of customer references?

15. Can we photograph your project?

          i.   Scheduled photo shoot for:

16. Open house?

17. 1-year warranty form sent to client:

18. Trigger Post-Project Follow-Up:

19. Trigger Seasons Emails:

Actionable Notes:

PRODUCTION PHASES

<u>Dates:</u> Phases:

Sitework
Demolition
Excavation
Footings
Foundation Walls
Drainage/Waterproofing
Backfill/Grading
Concrete slabs
_____**Phase 1**----------- Start of actual work to "ready for framing"

Framing
Roof structure/Roof/Skylights
_____**Phase 2**----------- Framing completely finished on all floors

Exterior doors/windows
Exterior trim/siding
Fireplaces/chimneys
Masonry
Patios (brick/stone)
_____**Phase 3**----------- Ext. trim completed (and all the above)

Decks/Porches/Railings
HVAC (rough-in)
Plumbing (rough-in)
Electrical (rough-in)
Security
Speakers/TV/Computers
Insulation
_____**Phase 4**---- Close-in inspection (HVAC, plumbing, elect., etc.)

Exterior prep & painting
Gutters/Downspouts
_____**Phase 5**----------- Ext. painting, gutters and downspouts

Drywall (primed)
Hardwood floors
_____**Phase 6**----------- Drywall and floors completed

Interior doors/trim/built-ins
Stairs/Railings
Interior prep & painting
_____**Phase 7**----------- Doors, built-ins and all painting prep work

All interior painting
_____**Phase 8**------------- All interior painting complete

Install kitchen cabinets/vanities
Appliances
Countertops
Tiles
Plumbing (Fixtures and trim out)
Electrical (Fixtures and trim out)
HVAC (Fixtures and trim out)
Bath accessories/shower doors/mirrors
_____**Phase 9**--------------- Install of all above

Carpet
Driveway
Landscaping
Trash/Clean up
Punch List
_____**Phase 10**-------------- COMPLETION OF PROJECT

CHANGE ORDER

No._____ ❏ Owner   ❏ Architect
                  ❏ Contractor ❏

NAME:_____   PROJECT:_____

ADDRESS:_____   INITIATION DATE:_____

_____   CONTRACT FOR:_____

TO (CONTRACTOR):_____   CONTRACT DATE_____

YOU ARE DIRECTED TO MAKE THE FOLLOWING CHANGE IN THIS CONTRACT:

_____

_____

_____

_____

_____

_____

_____

_____

_____

_____

_____

_____

The original Contract Sum was .... $_____

Net change by previous authorized Change Orders .... $_____

The Contract Sum prior to this Change Order .... $_____

The Contract Sum will be ❏ increased ❏ decreased ❏ unchanged by this Change Order .... $_____

The new Contract Sum including this Change Order will be .... $_____

The Contract Time will be ❏ increased ❏ decreased ❏ unchanged by (days) .... (   ) *days*

The Date of Substantial Completion as of the date of this Change Order is therefore .... _____

AUTHORIZATION

ARCHITECT_____   CONTRACTOR_____   OWNER_____

ADDRESS_____   ADDRESS_____   ADDRESS_____

_____   _____   _____

BY _____DATE_____   BY _____DATE_____   BY _____DATE_____

N

NOTE FROM THE EDITOR:

You might be thinking "I can do all of this on the computer or in the cloud. Why do it by hand?" Remember that FineCraft was established in 1985, before computers were in vogue and way before any estimating programs were readily available. This being said, the Estimate Sheets, Change Orders, Project Estimates, Job Estimates, Punch List, and the Production Phases can be used as PDFs for your tablet or laptop and be incorporated into your way of operating.

Your "Production Phases" may be in the form of a giant chart in a system such as BuilderTrend or CoConstruct, which is fabulous to help keep projects on schedule. The Production Phases, however, is intended to be printed and posted at job sites so the workers know where they're at in the project and what they have left to complete. A helpful tip is to enlarge it to poster-size and laminate it. Use dry erase markers to mark it up and make sure someone is on top of this so it's kept up to date.

Last but not least, if you're not using the cloud for files and whatnot, you should start doing so immediately. Sharing files is so much easier and your administration work becomes much more efficient because you can work from wherever you are, be it in the car or on vacation.

NOTE!

NOTES

CHAPTER 5
WORKING WITH HOMEOWNERS

CONSTRUCTION IS A MESSY BUSINESS. It's fundamentally dirty—dust and paint and mud and grime. But it can also be emotionally messy. Clients are emotional, and rightfully so. They're trusting us with their money (sometimes a lot of it!) to do work in their home. And as we all know, construction rarely goes 100% smoothly. It may fall behind, run over budget, or require changes on the fly.

As difficult as it can sometimes be to keep clients happy, the truth of the matter is that without clients, you have no business. Keeping your clients happy is one of the most important parts of running a successful contracting business. But it can also be the most challenging.

At FineCraft, we typically have about 15-20 projects going at one time. It's our job to make sure that each project is running as smoothly as possible and that each client feels like they're receiving our full and undivided attention.

We've worked hard at putting systems in place so clients feel taken care of—and it works. We have a 95 percent success rate with clients. When we finish a project, our clients feel like friends. They know the names of our crew members. We know the names of their kids and their pets and whether or not they mind if we play music while we're working.

This relationship is ultimately the reasons clients refer us to friends and family and come back to us when they need additional projects completed.

What's important is to begin building a solid relationship with clients from day one.

SET REALISTIC EXPECTATIONS

Clients always want things done fast and cheap. Unfortunately, construction is very rarely either of those things. So the most successful contractors can complete projects as quickly as possible and as cheaply as possible. A mark of a good contractor is helping clients understand that distinction.

The best thing you can do to create successful relationships with clients is to give them the most realistic expectations you can provide. Don't make promises about price or timing that are unrealistic. Be unflinchingly honest. If they end up paying more or waiting longer, it will reflect poorly on you.

This might mean that you lose some work to contractors who promise to do it more quickly and less expensively. I know that's a hard pill to swallow, but building a reputation as an honest and trustworthy contractor is much more valuable than the jobs you might lose to someone who gives false hope just to land a job.

Another important thing to remember here is that things change day-to-day in construction. There will be budget overages and delayed projects. Communicate this with the client as quickly as possible so their expectations are accurate and realistic at every stage of the project.

NOTE FROM THE EDITOR:

I met a colleague of George's who had this to say about George: "George is honest 100 percent of the time and that's both good and bad. People don't always want to hear the truth and sometimes they don't hire him because of it. But you always know you're getting the truth with him." I saw this first-hand as I observed George with his clients.

One couple, who were rehabbing a gorgeous condo in downtown Washington, D.C., told me, "George and his team are the ultimate problem solvers and we can always rely on them to be honest. This project has been delayed a lot—because we changed our minds a lot. George always told us exactly what to expect when we made those changes and we appreciate that."

That same day, that same client ran into a problem with the kitchen cabinet layout and as George held court with the clients and his crew, the client suggested a change—an expensive change. "Well," said George. "We can do that. But it's going to cost you another $5,000 and set us back two weeks." When the homeowner was clearly unhappy with that possibility, George shrugged his shoulders and said, "It is what it is." The client nodded his head in understanding, and they went back to the drawing board. They eventually landed on a solution that made everyone happy.

I saw this scene repeat over and over. The client would ask for something expensive/unrealistic/time consuming and George would shrug and say something to the effect of, "Here's what that means. Take it or leave it." There weren't any false promises, no reassurance that they could make it happen more quickly or less expensively. The clients didn't always love the answer, but it was clear that they expected this kind of transparency from him and appreciated always knowing where things stood.

SOMETIMES YOU HAVE TO TAKE A HIT

I never go into a project knowing I'll lose money. But invariably it happens. Sometimes things turn south because there was a mistake somewhere or something didn't work out the way you expected. You will lose money or won't make as much.

It's part of doing business. I don't drop projects because I'm losing money or because it's not turning out the way I wanted. As a business owner, it's my job to prepare for these unexpected downturns. Taking these hits every once in a while shows clients that we're willing to make their project successful no matter what it takes.

KNOW YOUR LIMITATIONS

I have very few clients who I wouldn't work with again, but there have been a few. We have a 95% success rate, but that other 5 percent left? I don't want to see them again in my life. It's how it is.

One client was particularly awful. Her project, which should have taken less than a year, stretched on to more than two. When I signed up for the project, I knew that she would be difficult, but I didn't know how difficult. It turned out that she was one of the most obnoxious clients I've ever had. Her husband was fine, but he didn't really say much to her.

So in one meeting with this couple (one of too many of the same meetings she called so she could sit down and start bitching), she was asking the same thing over and over. I acknowledged her concern at least four times.

After she repeated herself again, I said, "Let me tell you something. I've been doing this for 40 years, and I've learned some things. One being is that when somebody has the temperament that you have, you should first try to really work with that person, because maybe it isn't what I think it is. Then if it keeps continuing, I basically ignore it. You are an unenviable person and are one of the worst clients that I have met in my life. I don't want to talk to you again. If you want something, tell your husband, who will tell me. This is how we're going to do it from here on out."

She didn't handle it well. So I said, "Ever since we started, you have said nothing positive about the project. Not one thing. All you're doing is sitting here bitching about everything that is going on with your project. I'm not your punching bag. And if you want your punching bag to be your husband, well that's his choice."

Funny enough, her husband said, "Go on." We finished the meeting and as I was leaving, he pulled me aside and said, "You know, a couple guys at my office, they have some work and I gave them your name."

So I was willing to finish the project, but under my terms. I have the freedom to do that. Not because I'm the boss, but because I do everything for my clients. I really try to do the best job that I possibly can, and when they don't see that, I've had it.

Do your best to stick with even the most difficult clients for as long as you can. Not everyone is going to love you or be easy to work with. But you can deal with a lot to get the project done and keep everyone happy.

However, that does not mean you have to deal with abuse. You have every right to stick up for yourself and set boundaries and limitations with clients. Do so, and do so confidently.

I told that woman exactly what I thought of her and what I was willing to put up with. And they still hired us for another project.

MAKE SURE EVERYONE IS ON THE SAME PAGE

The best way to keep clients happy is to make sure everyone is on the same page. This goes along with being honest, but you should also have it written down. Set deadlines and do your best to stick to them. Keep the client updated about how the schedule is running and anything holding it up.

KNOW YOUR CREWS

It's obvious that you should only employ people who are trustworthy and competent. But this goes beyond just how they do the job.

At FineCraft, we work with crews who care for clients—not just the project, but the people. They're in the client's home for weeks or months at a time, so they have to be respectful of the client's space. They get to know the family so they need to be 100% respectful, all of the time. We want our crews to help clients with groceries, to clean up after themselves each day, etc.

NOTE FROM THE EDITOR:

I saw this in action multiple times. Once we were at a job site that was very much under construction. The homeowner came in for a meeting and took off her coat. She stood around looking for a place to lay it, but there was no furniture or any clean surface to rest it.

One of George's most senior and trusted crew leaders took the coat from her, then took off his own sweatshirt, lay it on top of a piece of drywall, then laid her coat on top of that. It doesn't sound like a big thing at all, but there was very clearly a level of respect and appreciation on all sides that was refreshing to see.

NOTE!

GET FACE-TIME

As the project moves forward, I make it a point to be on-site as much as I can. It helps clients feel like their project is a priority. It also gives me a chance to check on the project's progression.

TIPS FOR SUCCESS

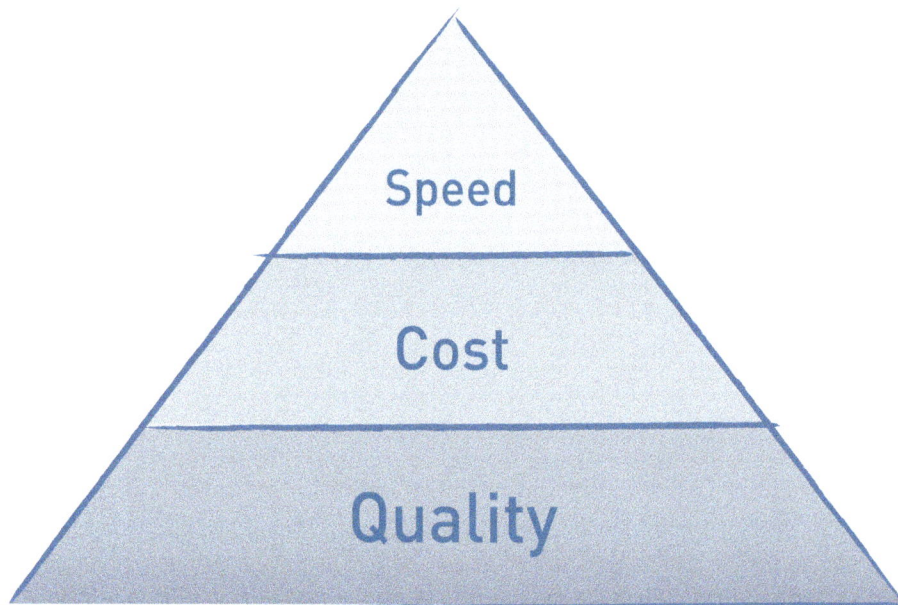

Speed

Cost

Quality

Customers can't have all three: But they can have two of the three.

If they want quality it will cost more money and may take longer. If they want it to be less costly they may get it faster and it will compromise quality. If they want speed then the cost will be higher and the quality may suffer. People say they want all three but it's not possible, so setting realistic expectations about the value of your work, the speed at which you work, and what it really costs to get all of the work done, is of utmost importance.

NOTES

CHAPTER 6

PRACTICAL SOLUTIONS FOR GROWING YOUR BUSINESS

IF YOU'RE ANYTHING LIKE EVERY OTHER BUSINESS OWNER in the world, growing your business is your top priority. You want to make money and in order to make money you have to get new clients.

As a contractor, you'll face a number of ups and downs as the market ebbs and flows. The best way to continue to grow your business in every type of market is to have a strong and consistent flow of leads.

Building a marketing strategy and a network of clients and colleagues who will regularly refer work to you takes time and attention. However, there are some things you can do NOW to get leads, nurture them, and turn them into clients!

GET A CONSISTENT FLOW OF LEADS

Leads are your bread and butter. In the past, many contractors relied on word of mouth to generate new business. That process still works really well—after all, when you do great work, clients will tell their friends and family.

But there are also a ton of other ways to generate new business that can have a huge impact on your growth.

BUILD A BETTER EMAIL LIST

An email list is a great way to not only gather and track leads, but also to nurture and share information with those individuals interested in your business.

- **Choose a system.** There are a number of free contact relationship management (or CRM) systems available that are user-friendly and offer email marketing automation. We suggest MailChimp or Active Campaign.

- **Import your leads.** Gather a list of client, colleague, and contractor emails and load them into whichever CRM you choose. You should also reach out to family and friends and ask for their email addresses.

- **Start sending!** We could write a whole book on email marketing alone, but the easiest way to ensure your email marketing is effective is to follow these rules:

  - Provide valuable information. Don't sell yourself all the time. Give important information about what to expect when working with a contractor, tips for homeowners, information about the local housing market, etc.

  - Reach out regularly (2-3x a month at first). You don't want to clog their inbox with multiple emails a week, but you should also reach out enough that they remember your name and look forward to hearing from you.

  - Personalize the information. Send separate emails to your list segments. Prospective clients want different information than old clients or the plumbers and electricians in your network. Send birthday and holiday emails.

KNOW THE POWER OF SOCIAL MEDIA

There are some social media best practices for contractors that will help you share your work and get leads from far and wide.

- **Choose the right platforms.** You can use almost any social media platform, but some work better for contractors than others. You want to show potential clients that your work is beautiful, high quality, and that your customers are happy. Facebook, Instagram, Pinterest and YouTube are perfect for sharing pictures and video and showcasing your work.

- **Post regularly.** Just like with email, you don't want to clog anyone's feed, but you should post regularly enough that your accounts have value and people start to recognize your name. Posting once a day is a great place to start.

- **Use hashtags.** Hashtags are the best way for people using social media to find exactly what they're looking for—but they have to make sense. For a small fee, services like Hashtagify will generate a list of the most popular hashtags in your industry. You can also study your competitors and other industry professionals. See which hashtags they use most frequently.

- **Get engaged.** Social media is all about engagement. If someone comments on your posts, be sure to respond quickly. Comment on and like other people's posts—anything to ensure your followers will see that your account is active.

- **Know the power of video.** Film clients who are seeing their project complete for the first time or film them talking about what it was like to work with you.

- **Engage.** If you have people commenting on your various social media platforms, you should always respond quickly. Answer questions, thank them for visiting, etc.

ASK CLIENTS FOR HELP

Your previous clients are your best source of reliable referrals. When they're happy with your work, they'll give you great reviews and refer their friends!

- **Survey your customers.** You can use surveys to gather great testimonials, but you can also ask clients about where they found you and what made them choose your company. This can help you focus your marketing efforts. You can also use their words for reviews on social media, in advertising, etc.

- **Give incentives for referrals.** If you're having trouble getting feedback, you can incentivize past clients to talk about your work. Hold giveaways if they share your posts on social media or refer friends for bids. You should also encourage them to post reviews on sites like Houzz, Yelp, Google Reviews, etc.

- **Talk to colleagues.** Ask the architects, plumbers, roofers, painters, everyone for referrals. They know the quality of your work and shouldn't have a problem referring you for jobs.

KNOW THE VALUE OF COLD-CALLING

Cold-calling can be awkward and frustrating—but it works. You don't just have to beg for work. The trick is to have a reason to call them other than just asking for jobs. Try to find an actionable step you can take or that you can ask them to take. Tell them about an open house you have coming up so they can see your work. Ask them if you can send them literature or set a meeting.

INVEST IN GOOD MATERIAL

We're not talking about quality cabinets. Having professionally designed and printed literature and marketing materials is about more than just a business card. Marketing material like brochures, flyers, and information packets give a great first impression and are tools just as important as hammers, squares, screwdrivers, and nails. They're something you can put in the hands of people you're talking to that showcase your product.

- **Hire professionals.** Hire a photographer to take pictures of your work. Hire a designer to lay out your marketing materials, using colors and fonts that match your branding. You can find professionals who get the job done but don't break the bank.

- **Hand it out to everyone.** Bring your literature with you everywhere you go and don't be afraid to pass it out! Canvass the neighborhoods you're already working in and those you'd like to work in, handing out literature and leaving it in mailboxes.

- **Build a good website.** Websites shouldn't just be pretty or have cool design elements. They should be interactive, dynamic, and easy to use. Have quality photos and client testimonials. It should be easy to navigate and have plenty of calls to action, like signing up for your email list, visiting an open house, or following you on social media. Make sure your contact information is easy to find.

- **Maintain a great website.** Keep your website updated, with a dynamic design. Make sure it has a clear path for your potential clients to come in and go straight on through to your contact form (see more about contact management below). After all, what you want is a reach. Get them to fill out the form or call you.

UNDERSTAND THE VALUE OF CENTRAL FILES

This is gold. Your hard-earned clients have experienced your blood, sweat, and tears it took to make their project perfect. Keep in touch with them. Use a contact management system like Infusionsoft, MailChimp, or Constant Contact.

Create email blasts that go out to clients - keeping a line out to old clients. Never lose touch with old clients while you make new ones. You never know when the old will need another project or who they will recommend.

AFFILIATE WITH LIKE-MINDED GROUPS

Join professional groups like the National Association Remodeling Industry, Associated General Contractors, the American Industry of Architects, local building industry associations, and more. Join the groups, make use of their resources, and attend meetings. You'll build professional relationships and learn more about the industry from other professionals.

You can also connect up with magazines and get on their email lists so you know about their events and what's happening in the industry. You can also take out ads, write articles, and more.

Surprisingly enough you can also reach out to builders that work in other service areas that you don't work in and tell them if they have potential leads that happen to be out of their service area that they send them over to you.

HIRE A PROFESSIONAL

We understand that finding leads and building your marketing strategy can be overwhelming and there is a lot to learn. As soon as you're able, hire someone to help you stay on top of all of that. This should be someone who knows how to manage lead generation and follow up on hot leads and can also manage your marketing initiatives.

THINK OUTSIDE THE BOX

We involved ourselves with a program that offers trips to cruises or resorts and this can be offered to your referrals and those who close with you. Check out destinationmotivation.com for details. Highly recommend.

We also hold open houses three or four times a year to attract interest. Potential clients can come in and look at a completed project and ask us questions about the process. In order to generate interest, we mail out to our central files and canvas the neighborhood inviting all the neighbors.

WHAT HAPPENS WHEN THE LEADS DRY UP?

It's our worst nightmare—all of a sudden we're finishing up a project and we have no new leads. It can be panic-inducing. But if you put the above suggestions into action you should have a steady source of leads coming in.

A very important thing to remember when your business is drying up—pour money into promotion and marketing. You have to be known. Spend money on door hangers and go canvas a neighborhood. Make sure your projects have yard signs and info boxes with your literature. Make sure your trucks are promoting your company and services. Get involved with EDDM (Every Door Direct Mail). This is an inexpensive way to get out postcards.

The bottom line is that when it comes to building a consistent flow of leads you have to keep information flowing out and never stop. Post, email, call, visit consistently. If you put all of this in place, the leads will never stop flowing!

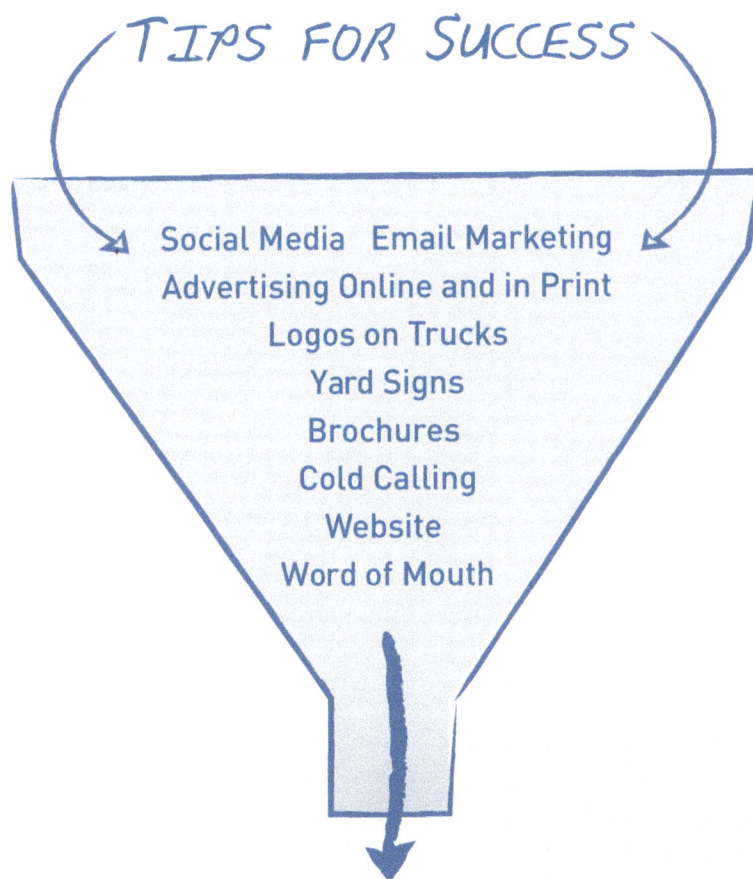

TIPS FOR SUCCESS

Social Media   Email Marketing
Advertising Online and in Print
Logos on Trucks
Yard Signs
Brochures
Cold Calling
Website
Word of Mouth

All of these actions will result in increased leads—which means more sales!

NOTES

CHAPTER 7
SURVIVING THE ROLLER COASTER

ANYONE WHO HAS BEEN IN THE CONSTRUCTION BUSINESS for any length of time will tell you that the construction market fluctuates. It's a normal part of business. It can definitely be stressful during the times that there doesn't seem to be any work or people want you to complete projects for pennies. But if you plan ahead and prepare for the lean times accordingly you can not only make it through the lows, but also come out on top.

Most everyone knows about the big market downturn in 2008. "The Great Recession" as it's now known, had a dramatic impact on almost every industry, but it hit construction particularly hard.

UNFORTUNATELY, I WAS COMPLETELY UNPREPARED. I DIDN'T SEE IT COMING AND WE ALMOST WENT UNDER.

When I first realized that the market was crashing, I started getting any project at all by simply underbidding everyone else. I just offered cutthroat pricing and ended up doing the job at cost. But that model was not sustainable so I realized I had to come up with a better solution.

I realized that eventually, the market would turn around. I would still need good people on my team. The way I saw it, I had two options: I could lay off my people, and then when things turned around, I would be completely without anybody because I got rid of all of my people. I'd have to start all over again. And in reality, if I laid them off their chances of finding work again would be minimal and they wouldn't have any work at all. The second option was to find a way to keep them around without me losing every penny I made.

So I made a strategic decision and made an agreement with my guys. I told them that I would give them as much work as I could but I wouldn't be able to pay them their regular salary. I told them I would owe them. They would continue to work for us, but they may not get paid all the money right away.

We basically reduced payments and put everybody on half-pay. I told them that when times got better again, we would pay off the balance. And that's how we did it. All but two of my guys agreed and the rest are still with me today.

If I may say so, it was brilliant on my part, and it was even smarter on their end to accept. They realized that they could have left to try and find other work but might have found nothing or find something where they were making less than my half-pay option. When things turned around, they were bumped back up to their regular pay and I paid them back every penny I owed from when things were lean. We all survived it.

So I'm very proud of that decision. But that doesn't mean I've never made a mistake. For me, the 2008 downturn was a double whammy. Not only did we get hit by the recession, but prior to when the downturn began, I left the company for about six months. I was in contact by phone and so forth, but I was not really hands-on. I left a manager in my stead. A manager who ended up letting me down.

We had several projects we had signed up for and I was anticipating the company making about three or four hundred thousand dollars profit. I returned from my absence to discover that not only were we not going to make three hundred thousand dollars profit, but we were in the hole for $200,000. So on top of the recession, I had to come out of the debt he put me under.

As it turned out, he had stolen part of the money, but more than that, he was so incompetent that he basically just let everything take its own course. He let all the guys continue without any guidance or any kind of leadership. Money gets eaten up incredibly fast when you have twenty people doing nothing and getting paid.

That man died before I could fire him and I was left to clean up the mess. But we got through it and are doing better than ever. But know who you're hiring and know who you're leaving in important positions. Make sure they are honest, reliable and competent.

There's a lot for other contractors to learn from my mistakes. There are ways to put procedures in place to safeguard your business from downturns or other unexpected problems.

First of all, it's vital that you're honest with the people you're working with. Tell them the truth and give them an accurate representation of what's going on with the business. When there's a problem, try to find a solution that works for everyone.

Next, and as I mentioned before, never sacrifice marketing dollars, even when things are lean. It might not feel smart to spend money you don't have on promotional materials or advertisement, but if you don't, business will dry up 100 percent, with no processes in place to generate new business.

THERE'S A LOT FOR OTHER CONTRACTORS TO LEARN FROM MY MISTAKES.

After the 2008 fiasco, my son Niko stepped in to lead marketing initiatives for FineCraft. It started as a temporary solution to address everything that was happening after the downturn. But it turns out he likes it and he's good at it.

Niko's temperament, education, and energy level has really pumped a lot of energy into the company. It has its minus in the sense that he drives me nuts. He is a slave driver—I call him a pest. Half-jokingly.

But more important than all of that, he has taken an aspect of work that I could not also do on top of everything else I was doing. More people are finding out what FineCraft is all about and want to use us to do their projects. We came out of the downturn stronger than ever. That wouldn't have happened if I hadn't been willing to let him step in and start spending precious dollars and time on marketing.

The ultimate safeguard is to approach developing your organization as an enterprise. I think a big problem is that 90 percent of contractors look at themselves as a technician only—good only at building roofs and walls and doors. And they may be very strong as a technician, but they don't look at themselves as an enterprise, as an organization.

IF YOU LOOK AT AN ORGANIZATION AS A WHOLE, IT HAS SEVERAL PARTS.

Each of those parts must be properly addressed and taken care of. We'll break them down here.

- You should have all of your legal rudiments in place, like licensing, insurance, workers compensation, etc.

- You should have all the necessary personnel in place and train them adequately.

- You have to have materials, records, and bookkeeping in place to track every element of your business.

- You need effective promotions, marketing, and advertising in place so you get the word out about your business. This includes public relations, in order to have a good name in the community.

- You must have quality control, processes for making corrections, dealing with customers, etc.

WHEN ALL OF THAT IS IN PLACE, THEN YOU GO INTO THE TECHNICAL ASPECTS OF IT.

If an individual goes out and tries to start a business, using only his own work and skills, and doesn't really put attention in any of these things, he's going to starve. Because when it's feast time and business is booming, he's going to work himself to death, because he's only himself and a couple carpenters. He's not going to have time to do the other stuff he needs to do, because he's too busy actually building. He's too busy doing things that could be completed by a subcontractor or contractor. And when times are lean, if he does not put attention on these other things, chasing after any dollar he can get his hands on, he will have nothing in place to sustain the business for any serious length of time.

MY BIGGEST MISTAKE IN MY BUSINESS WAS NOT PUTTING THESE PRACTICES IN PLACE SOONER.

The solution to long-term survival is to somehow find a way to work with and put attention on all of these elements I'm talking about. How will you find more work? How will you train people? How will you bring new people in? How will you correct your mistakes? These things, a little at a time, should be an integral part of your everyday living.

As you become better at these things, and as you put some of these things in your company you'll start feeling better, and better, and better. That is the long-term solution to coping with any kind of up or down in the industry.

My biggest mistake in my business was not putting these practices in place sooner. For a very long time, I was stuck on being the only one. Because I was so good at what I was doing, it was exacerbating my feeling that I could handle it all.

But the fact is that I can train somebody to be like me. And if I can train someone to do something in an hour instead of half a day, then I have something, don't I? Then I don't have to be the only one. Then, when they get better than me, God forbid, then I have really done something.

It freed me up to go and do what I wanted. I was able to go and grow the business, I was able to spend time with my family, and all areas of my life improved in a million ways.

When I was doing everything myself, it was literally a disaster. I didn't really have time for anything. I didn't have time for my kids, I didn't have time for my wife, I didn't have time for anybody. I was just working my butt off. And what did I have to show for it at the end of the day? Just enough money to make a living.

In this business, if you don't focus on each important aspect of your business and take the time to learn how to put all essential processes in place, you will not grow. Yes, there will be times you can make tremendous money, but you have to keep on doing it. But it's impossible to do it alone.

The bottom line is that I believe if things never go wrong, they can never go right. You can learn from my mistakes and take notes about what works for me. But no matter what, you will make mistakes. There will be hard times. But learn from your own mistakes and make the changes necessary to do better next time.

NOTE FROM THE EDITOR:

When George told me about the information contained in this chapter, it was refreshing to hear how matter of fact he was about the mistakes he made. He told me that out of all the difficult times he's had in the business and things he's done wrong are the reasons he wants to share what he knows with other contractors.

NOTE!

TIPS FOR SUCCESS

On the next page, we've gathered a list of all of the "boring stuff" you need to have in place in order to survive downturns or unexpected life and business events.

THE MASTER ADMINISTRATIVE LIST

You can task your employees with these items or hire experts, just make sure you check every box! Also, remember we're not experts in all of this, so be sure to check with your trusted attorney, financial advisor, and accountant so they can help you make the right decisions for you and your business!

IT'S NOT FUN, BUT IT'S VITALLY IMPORTANT!

☐ Insurance

- Including (but not limited to): health insurance for you and your employees, worker's compensation, and perhaps an umbrella policy for extra protection.

☐ Life Insurance

- For YOU and any business partners

- Possibly for employees

☐ Bookkeeping

- Managing payroll

- Tracking income and expenditures

☐ Accounting

- Managing taxes (quarterly and yearly)

☐ Investment Advisor

- Can help you wisely invest your money and save for the future

- Can advise you on any savings/retirement plans you might want to offer employees

☐ Licensing

☐ Marketing Pipeline

☐ Sales Pipeline

NOTES

CHAPTER 8
ADMINISTRATION AND ORGANIZATION

AS WE SPOKE ABOUT IN CHAPTER 7, there are certain divisions of your business that have to be properly set up and maintained in order to keep your business running smoothly, thriving when things are good, and sustaining when things are lean.

In order to do this, you need to have the proper administration and organization elements in place so that you can get these processes in place and focus on growing the business.

For FineCraft, I've built a capable team comprised of competent individuals who know what tasks they're responsible for, what I expect of them, and how they fit into the team as a whole.

DASHA is a Project Coordinator. She handles the contractors that work for FineCraft. She also ensures communication is flowing to and from FineCraft and that the customers are always in the know. She ensures the project is flowing well and that the customer is satisfied. She also requests payments when they are due. She's the middleman that handles the administration and timeline of each project and that the quality is kept up to our standards. Dasha uses routing forms, which are essentially checklists detailing who is responsible for what and how it can best get done in a certain sequence.

NIKO is our Business Manager. He handles all legal, business back end and finances, as well as all marketing and promotion of the company and warranty tasks that come up. He ensures money in and out is controlled, that people are paid, and that the company stays solvent. His marketing and promotional tasks include literature, videos, online presence, networking events, affiliations, and more.

PATRICK is our Sales Representative. He does whatever is necessary to sell the job. If this means going to a homeowner's house at 8 pm on a Sunday to answer any questions, he's sure to be there to do just that because it's the care that counts. Being a salesman is about caring for the potential buyer. People get this idea that they're like a used car salesman but it's just not true.

The real salesman sells by providing what is needed and showing the features the company can offer that will benefit the buyer by catering to their needs. In order to show what we can bring to the table, what we have found to be very helpful is showing our buyers work that we've done. References are great but showing our work is the cherry on top. One can't get any closer to our work other than seeing it first hand and even talking to the homeowners.

OMAR is FineCraft's Bookkeeper and Materials In-Charge. He tracks all receipts and is the actual check writer. He buys material for jobs. This is a bit unconventional, but it's worked for us for up to $4 million annually. When we break that benchmark then we'll have to get him an assistant.

VIVIAN is my Assistant. Her work is similar to Dasha's, but she helps me so I can be on the job and work as needed. She's on the road with me every day, meeting with clients, running errands, and keeping me in check!

AND THEN THERE'S ME! I'm the technical expert who ensures all quality is up to our standards. I'm on-site as much as possible meeting with clients, putting out fires, and helping move projects along. I also handle estimates for projects.

TIPS FOR SUCCESS

We've included a blank Income and Expenses Tracking Template and a Blank Contract Template to help you get started on organizing some of your most important administrative tasks. But remember—we're not attorneys or accountants, so make sure you consult with a professional!

- CONTRACT -

Work to be performed at:

Name:
Address:

Phone Number:
Email:
Date:

We hereby propose to furnish the materials and perform the necessary labor to complete the following work:

Description of work to be performed

As described in proposal herein:

All material is guaranteed to be as specified, and the above work will be performed in accordance with the specifications provided for the above work and will be completed in a substantial workmanlike manner for the agreed sum of $.

Payment Plan:

$CONTRACT TOTAL

*WILL THIS CLIENT USE OUR FINANCING?

NOTE: Use of our financing carries an additional 3% processing fee.

PERMITS:

It is the responsibility of the Homeowner to pull any building permits relating to their project before start date of project. _____ (INITIALS)

ALLOWANCE:

In construction, an allowance is an amount specified and included in the construction contract (or specifications) for a certain item of work (e.g., appliances, lighting, etc.) whose details are not yet determined at the time of contracting.

CHANGE ORDERS:

A change order is a change from the base contract that adds or deletes and is agreed between the Homeowner and the contactor (BUSINESS NAME). The contractor shall provide a written change order to be signed by the owners prior to starting work. Change order payments are due to BUSINESS NAME at the time the change order agreement is signed by the Homeowner.
_____ (INITIALS)

SUBSTANTIAL COMPLETION:

The project has been completed to a point where it is usable for the purposes it was constructed and has no major unfinished items. Only minor, non-essential items remain which are cover under Punch List (see below). _____ (INITIALS)

PUNCH LIST:

Punch List is only done ONCE with owner, contractor and architect (if available). The purpose of this Punch List is to complete miscellaneous, minor items found to be incomplete in an effort to terminate the contracted work. The Punch List is to be signed and dated by both parties after the Punch List Walkthrough and after the Punch List is found to be completed by both parties. Upon completion and signing off of ALL Punch List items, owner releases final payment to BUSINESS NAME Any further items found after completion of Punch List are still BUSINESS NAME responsibility which will be addressed under Warranty work and will be addressed in a timely manner. _____ (INITIALS)

FINAL COMPLETION:

A 100%, fully completed project – including all punch list items. Finishing the Punch List equates to a Final Completion of the project.

FINAL PAYMENT:

Final Payment is due upon signing off on Punch List signed by both the homeowner and BUSINESS NAME

HIDDEN CONDITIONS:

The contractor is not responsible for extra work resulting from hidden existing conditions that may be discovered during work. In the event that hazardous materials are discovered, necessary remediation work and costs are not included in this agreement. It is the responsibility of the homeowner to advise BUSINESS NAME should they suspect any hazardous materials were used on their home at any time in the past. Any hidden conditions constituting a change in design or cost must be brought to the attention of the owner immediately. For changes, refer to CHANGE ORDERS above. _____(INITIALS)

WARRANTY & INSURANCES:

BUSINESS NAME guarantees the labor and materials as specified. All labor warranties shall be one year from Substantial Completion of the contract. The labor warranty constitutes the repair of any of the products provided that their failure is due to faulty workmanship. The homeowners are not responsible in any way for injuries or damages to any party or representative of BUSINESS NAME's team. The contractor will make every effort to keep the work site safe but is not responsible for injuries incurred by others on the site who aren't under our insurance coverage. Any material ordered by the homeowner must be approved by the contractor. Contractor is not responsible for order placed by the homeowners without the contractor's consent. Contact customercare@BUSINESS NAMEcontractors.com , should the need arise for a warranty item. _____ (INITIALS)

NON-PAYMENT:

If the scheduled payment is 10 days past due, the contractor reserves the right to stop work on the project within 5 days of written notice. The owner is responsible for all expenses related to non-payment, including reasonable legal fees, payment for work already performed, and other losses.

DISPUTE RESOLUTION:

In the event of a dispute that is not resolved between the contractor and owner, both parties agree to mediation prior to litigation as a term of this agreement.

SCOPE OF AGREEMENT:

This contract reflects the entire agreement between the contractor and the owner, and takes precedence over any and all previous written or oral agreements.

BUYER'S RIGHT TO CANCEL:

If this agreement was solicited for your residence and you no longer want the goods or services, you may cancel this agreement by mailing a notice to BUSINESS NAME The notice must say that you no longer want the goods or services and must be mailed before midnight of the third business day after you sign this contract. The notice must be mailed to: BUSINESS NAME 104 Summit Hall Rd Gaithersburg, MD 20877. If you cancel, BUSINESS NAME will refund your full cash down payment. If you cancel after the Buyer's Right to Cancel period of 3 days you forfeit your right to the refund of the down payment.

CONDITIONS:

- Our service in this contract does not include formal advice or service with selections with tile, countertops, cabinets, paint colors, lighting, built-in design or built-ins, etc. We informally help with all this, but if you'd like a designated designer to assist with such services we can offer our affiliated designers/decorators to service you for an additional fee.
- The estimated time for completing the base project is approximately_____
- Delays not caused by BUSINESS NAME (weather, inspections, special materials, change orders, etc.) will have to be added to the completion date as necessary.
- Before starting work BUSINESS NAME will provide a copy of our license and insurance to homeowner upon request.
- Building permits to be provided by Homeowner if they are required unless otherwise agreed.
- All inspections will be coordinated by BUSINESS NAME as needed.
- During construction, electrical and water utilities shall be provide by the Homeowner.
- Homeowner-provided materials need to be delivered to the work site. If Homeowner chooses, BUSINESS NAME will pick up the Homeowner's materials for a delivery and service fee if necessary.
- BUSINESS NAME is not responsible for any delays caused by delayed selections by homeowner-provided materials.

OTHER CONDITIONS: *(any 'other conditions' should be written here in writing during Contract Signing.)*

The following are not part of the contract unless specifically included in "Description of Work to be Performed" section:

- Painting int. and ext. other than what's included in Scope of Work
- Carpet
- A/V work & telephone wiring
- Security systems
- Landscaping
- Electrical heavy-up
- New water/sewer line
- Fees or surcharges by utility companies
- Inspection fees if any
- Wall check surveying or engineering fees

APPROXIMATE START DATE: **APPROXIMATE COMPLETION DATE:**

_____   _____

The foregoing terms, specifications and conditions are satisfactory and hereby agreed to. You are authorized to work as specified and payment will be made as outlined above. Upon signing this agreement, the Homeowner represents and warrants that he or she is the owner or the authorized agent of the aforesaid premises and that he or she has read this agreement. This contract may be withdrawn if not accepted by the Homeowner within 10 days if work has not commenced.

BUSINESS NAME ACCEPTANCE:

Signature:_____Date: _____

Print: _____

HOMEOWNER ACCEPTANCE:

Signature: _____ Date: _____

Print: _____

Existing Projects Income	INCOME	Prd.1:	Prd.2:	Dates	Notes
1					
2					
3					
4					
5					
6					
7					
8					
9					
10					
11 Change Orders factored into PLANNED INCOME					
12					
13					
14					
15					
16					
17 PLANNED INCOME					
18 NEW Projects Income	INCOME	Prd.1:	Prd.2:	Dates	Notes
19					
20					
21					
22					
23					
24 *Income(not factored into operation unless otherwise noted)*				BONUS income	
25 Expense: Office	Expense	Prd.1:	Prd.2:	Dates	Notes
26					
27					
28					
29					
30					
31					
32					
33					
34					
35					
36					
37					
38					
39					
40					

_____ [Month]   Prd. 1: 15th   Prd. 2: 30th

	Expense	Prd.1:	Prd.2:	Dates	Notes
41 Expense: Office					
42					
43 Expense: Insurance	Expense	Prd.1:	Prd.2:	Dates	Notes
44					
45					
46					
47					
48					
49 Office Expenses Subtotal					
50					
51 Expense: Materials	Expense				
52 Misc. material expenses					general material (to be confirmed)
53					
54 Subtotal					
55					
56 Payroll: Subtotal					2 payrolls this month
57					
58 Other Expenses:					
59 Promo					promo for the month
60					
61					
62 Subtotal					
63 Subcontractors & Vendros (no debt)					
64					
65					
66					
67					
68					
69					
70					
71					
72					
73					
74					
75					
76					
77					
78					
79					
80					
81					

	Expense	Prd.1:	Prd.2:	Dates	Notes
82					
83 Subcontractors & Vendros (no debt)					
84					
85					
86					
87					
88					
89					
90					
91					
92					
93					
94					
95					
96					
97					
98					
99					
100					
101					
102					
103 *Include additional vendors & expenses that come up during month*					
104					
105					
106					
107					
108					
109					
110 Subtotal					
111 DEBT	Expense	Prd.1:	Prd.2:	Dates	Notes
112					
113					
114					
115					
116					
117 Debt Total:					
118 Subtotal					
119 Expenses for the period					
120 Income/Expenses Difference					
121					
122 REPORT:					
123					

3

NOTE FROM THE EDITOR:

We use Quickbooks to manage our income and expenses, to see our profit and loss, to gauge how each project is progressing financially, but we also have this monthly Income and Expenses Tracking Template to help us with our monthly planning. This includes projects that are in projects, projects we plan on closing, office expenses, vendors and subcontractors, etc.

This is then broken down further into weekly templates which drawn from the monthly. Over the year this key activity has proven successful. The construction business is a volatile industry and one can't ever be too cautious and in control with the finances.

NOTE!

NOTES

CHAPTER 9
FOCUSING ON THE FUTURE: PREPARING FOR RETIREMENT

AS A BUSINESS OWNER, retirement is probably one of the last things on your mind. You're trying to close that next big sale or simply figure out how to cover payroll. On top of that, if you truly love what you do, you might think you'll never retire. The truth is that we don't always get to make that decision. Sometimes unexpected events happen to ourselves or a loved one, such as an illness, accident, disability, that dictates that decision for us.

It is vital that you have a plan for a time that you may no longer be running your business. This might be a plan for how you will generate cash flow in retirement to maintain your lifestyle or how you will turn that business over to someone else to run in your absence.

Unfortunately, business owners are frequently on their own to figure out how to plan for their financial future and the long-term financial prosperity of their business. They might have an accountant, but he or she only helps with current and past taxes. Or they might have a financial planner who's helping them plan for their personal future—not the future of their business.

To make matters worse, financial professionals are often more interested in selling you a product than they are in helping you create a long-term, comprehensive plan.

Think of it like a builder who is more interested in selling fixtures and appliances than he is in helping you build your dream house. This is why it's important to do your due diligence before hiring a financial professional.

BEFORE WE DISCUSS HOW TO FIND THE RIGHT *FINANCIAL GENERAL CONTRACTOR* TO BUILD YOUR FINANCIAL FUTURE, LET'S TOUCH ON WHAT GOES INTO A SUSTAINABLE FINANCIAL PLAN.

As the great Stephen Covey says, "Begin with the end in mind." Envision what the ideal retirement would look like for you. Does it involve lots of time traveling, volunteering, time with family, or starting a new project? Most retirees get bored rather quickly once the novelty of not having to get up to an alarm clock wears off. So have a plan. Consider writing your memoires or mentoring others in your craft. Turn a hobby into a side-gig. Find a place to volunteer that aligns with what's important with you. Being productive and active in retirement is just as important as having the cash flow to maintain your lifestyle.

You have spent most of your life creating money from time, in the form of work. With a well-designed retirement plan you will be able to reverse those flows and begin creating time from money. Quite literally you will be able to buy yourself (free) time with the money you have accumulated in your retirement plan. If you do not begin planning today in order to fund tomorrow, you will have to continue putting time in to get money (keep working), rather than money in to get time (financial freedom).

BUILDING YOUR FISCAL HOUSE

From my perspective, there is a sequence to building a sustainable plan and it starts with a blueprint. Much like building a house, you will create a plan before you ever begin to break ground.

Once you do begin the actual building process, the first section of your financial house is the foundation. This is your most secure assets or income that are likely to not change in the future. They are income sources that are principal protected and can be counted on to be there when you need them. They could include your Social Security, pension (if you have one), government bonds, or fixed annuities. This doesn't include variable annuities because although they may or may not generate a guaranteed income source, they are generally not principal protected.

Secondly, you need to frame the walls of your new home. Municipal or corporate bonds, real estate (investment property or real estate investment trusts – not personal residence), alternative investments such as master limited partnerships, private equity, hedge funds, and so on, are investments that may be considered for this portion of your plan. These investments are considered a step up in the risk scale from your principle protected accounts.

Finally, you would put on the roof. This is the portion of your home that traditionally holds the most risk (think wind, sun, and rain) and is also the one that will likely be more actively managed (or replaced) more frequently than the previous two. The roof includes investments such as stocks (equities), mutual funds (MF), exchange traded funds (ETF), and variable annuities. This is where most of your risk will likely be in your plan.

THE LAST THING YOU WANT TO DO IS RISK YOUR ENTIRE NEST EGG.

Keep in mind, however, that there are various levels of risk even within these investments (i.e. stocks, bonds). For example, you could be in a very risky "junk bond" or a more conservative preferred stock. All securities are not equal and you will need to do your homework or hire a professional to assist you with this. The last thing you want to do is risk your entire nest egg a few years before you'll need it for retirement simply because you were taking too much risk. I think most individuals find their lifestyle would change more in retirement if they lost half their money than if they doubled it. So why take on too much risk if the added growth would do less for your lifestyle then the potential loss?

John Bogle, the founder of Vanguard, often cites the rule of 100, which is a rule of thumb to help you determine how much of your money should be at risk at any given time of your life. The rule recommends that you should subtract 100 from your current age to determine the maximum percentage you should have invested in risk-based assets such as stocks, or as we say, in the roof. For example, if you're 40 years old it is recommended that you could have as much as 60% of your portfolio in the roof. If you're 60 years old you would typically want no more than 40% in the roof, and so on. Of course, it's different for individual situations but it is a good general guideline that can be used as a starting point for determining your risk tolerance.

PUTTING A FENCE AROUND YOUR HOUSE

Now that you have your dream home built, it's time to put a fence around your property. Think of the fence as the resources that help put a perimeter around your home. These often include your legal documents as well as various types of insurance such as life, health, disability, and property and casualty.

Legal documents in this context are your estate planning documents. These include a will, living will, health care directives, power of attorney, and possibly a trust. It's important to have these documents created for yourself as well as for the owners of your business. Traditionally you will want to have them reviewed and potentially updated every three or four years, or as circumstances change, so they stay current with legislation as well as any changes that may have occurred in your life.

Most of these forms will actually help you while you are living, so don't think of them as simply documents to use after your passing. I also recommend getting these documents done in your resident state and by an attorney who frequently works with this type of planning. It wouldn't be ideal to hire your cousin, who's a corporate attorney and does wills on the side. Hire someone who's experienced, even if it is slightly more expensive.

Another important part of your fence is life insurance, especially for business owners and high net worth individuals. Upon the passing away of the owner, in addition to helping with the cost of final expenses, often more importantly it can be used to replace future lost income, pay off debt, buy out a partner, or cover taxes that may be due on the estate or the business.

A colleague of mine was once making a proposal to a business owner to purchase life insurance. He proposed he buy enough life insurance to help cover the cost of the potential estate taxes due upon his death (keep in mind that the IRS only gives you nine months to file and pay estate taxes after a death). He happened to own the largest auto dealership of its kind in the tristate area and had three sons who worked for him, running the business. Their plan was for the sons to eventually take over their father's company when he retired. The father told the life insurance agent, "I'm not going to write a big check to an insurance company simply to avoid writing a big check to the IRS." Needless to say, he did not buy the life insurance.

Tragically, less than six months later he was killed in an auto accident with this wife. Neither the sons nor the

HIRE SOMEONE WHO'S EXPERIENCED, EVEN IF IT IS SLIGHTLY MORE EXPENSIVE

business had enough cash available to pay the taxes owed by the IRS deadline due to the large value of his estate. They were forced to sell the business at a loss to their largest competitor and now the three sons are employees of the new firm. I'm sure you don't want this to happen within your own business after all the blood, sweat, and tears you have invested in it.

Another benefit that many life insurance policies have today is living benefits. These are benefits you can potentially receive while you are still living. These include taking all or part of the death benefit tax-free for the purpose of long-term care, if the need arises. There are also ways you can access a portion of your death benefit in the form of tax-free income through loans against the death benefit. Every policy is different and often there are costs associated with this strategy so you will need to do your homework to determine if this is a potential option for you.

This strategy is similar to a reverse mortgage, where you tap into the equity of your home through tax-free income. Only with life insurance, you are traditionally buying that equity with pennies on the dollar through a death benefit. When you pass away, the insurance company pays back the loan from the death benefit and your heirs get the balance of the death benefit in the form of a tax-free cash payment.

FINDING THE RIGHT ADVICE-GIVERS

Just like when you're building an actual house, not all contractors or suppliers are the same. It's important that you take the time to interview the individuals and firms you'll be working with to help build your financial future.

There are various types of advisors and specialty advisors. Some are only licensed to sell securities, while others are only licensed to sell insurance, while others still are more diverse in their approach and offer both securities as well as insurance.

When I was growing up, my mom always used to tell me to watch who I hung around because they would have a great impact on my success or failure. The same is true when it comes to finding employees or business professionals. Who you surround yourself with will often have a greater impact on your success than many other factors.

When you are interviewing a financial professional you will likely want to work with someone who you not only feel comfortable with but one you feel listens and understands your needs. This person will hopefully be your financial advocate to guide your future decisions and help ensure you meet the financial goals you have established for yourself, your family, and your business.

You can begin by finding out if they are a fiduciary. This is determined by *WHO YOU SURROUND YOURSELF WITH WILL OFTEN HAVE A GREATER IMPACT ON YOUR SUCCESS THAN MANY OTHER FACTORS.* their license type. Fiduciaries are known as investment advisors and their licenses can be verified at www.advisorinfo.sec.gov. They traditionally get paid a flat fee, hourly rate, or a percentage of the assets they manage. The second type of financial advisor is regulated under FINRA and their license can be verified at www.brokercheck.finra. org. These advisors traditionally get paid a commission for the transaction of buying or selling a security on your behalf.

Finally, there are insurance agents. They are regulated on the state level, so you will need to contact your local department of financial services to verify their license and the carriers they are appointed to represent. These agents are also compensated through a commission tied to the product they are selling.

You will need to decide which of these advisors, or a combination thereof, are best for you. Finding the right advisor to help you build your financial plan may seem daunting, but from my perspective it is one of the most important decisions you will make in the process. Think of it like this: you may have the best quality materials in the world to build your house, but if you hire a poor-quality contractor, you'll end up with a poor-quality house.

I know some of this can be confusing and overwhelming. But you've worked way too hard to not take planning for your future seriously! Take the time and money necessary to have a thorough plan in place to take of your business, your future, and the people you care about.

NOTE FROM THE EDITOR:

This section was written with help from the amazing Celine J. Pastore, CRPC®, RFC®.

You can reach Celine at:

SimplePath Retirement, LLC
3060 Palm Harbor Blvd., Palm Harbor FL 34683
(727) 304-6000
Celine@SimplePathRetirement.com

NOTE!

NOTES

DOWNLOAD THE PDFS

You can download all of the worksheets in this book! Just visit the website below and we'll send them right to your inbox.

https://la351-d04315.pages.infusionsoft.net

FINAL WORDS

This book is the tip of the iceberg when it comes to establishing a successful company, but it contains fundamental and actionable information for you to win and to succeed in this industry. This book was written for you - the contractor. I want you to succeed. If you have questions about what I've covered feel free to reach me at: customercare@finecraftcontractors.com.

BE ON THE LOOKOUT FOR MORE TO COME!

NOTES

NOTES

NOTES

NOTES